FROM BREATH TO BELONGING

Listen as the Earth Speaks

AF661075

Wing Commander Srikanth Balagandar
Indrajit Shaw

Vitasta

Published by
Renu Kaul Verma
Vitasta Publishing Pvt Ltd
4348/4C, Ansari Road, Daryaganj
New Delhi - 110 002
info@vitastapublishing.com

ISBN: 978-93-47098-89-5
© Wing Commander Srikanth Balagandar & Indrajit Shaw
First Edition 2026

MRP ₹495

All Rights Reserved.
No part of this publication may be reproduced, stored in a retrieval system, or transmitted in any form, or by any means–electronic, mechanical, photocopying, recording or otherwise–without the prior permission of the publisher. Opinions expressed in this book are the authors' own. The publisher is in no way responsible for these.

Edited by Priya Khanna
Cover Design by Shubhpreet Kaur
Layout by Rohit Gautam
Printed by Vikas Computer and Printers, New Delhi

Contents

Advance Praise

Terrain forms an important component of military planning. Yet, we rarely pause to consider a whole new dimension of the 'Earth' that sustains the living being. *From Breath to Belonging* is a refreshingly new exploration of this vital attribute It is a soul stirring tribute to the quiet wisdom of nature. It draws you closer to the pulse of the planet inviting reflection, reverence and sense of connection. Wg Cdr Srikanth Balagandar & Indrajit Shaw approach climate change not as an academic or activist but as an inspiring and concerned sentinel of society. Their admirable articulation is hinged on meticulous research and passionate conviction.

Major General Suresh Mohanty (Retd)

Defence Strategist and Commentator on International Relations

Serving in the serene yet fragile ecosystems of Arunachal Pradesh has taught me that protecting the environment is as vital as guarding the frontiers. *From Breath to Belonging* captures this ethos with rare insight. It bridges military precision, modernity with spiritual reflection, climate urgency with cultural pride. Srikanth and Indrajit as buddies, their writing is evocative and rooted, reminding us that the Earth speaks sometimes in legends, sometimes in data, always in truth. This book is not just a read; it is a reflection on responsibility.

Col Anand Murty

Indian Army

For over two decades, I've been working with waste—not just transforming it into products, but redefining how we perceive it. To me, waste is never the end; it's always the beginning of something meaningful. That's exactly what I felt while reading *From Breath to Belonging.* This book doesn't preach sustainability—it lives it.

The author duo, Srikanth & Indrajit have beautifully captured the quiet power of nature and the deep-rooted wisdom we often overlook. It's not just about green practices—it's about listening, observing, and returning to what we were always meant to be: connected.

This book feels like a parallel journey to my own—where every discarded petal, every forgotten byproduct, holds purpose. *From Breath to Belonging* is a timely reminder that sustainability isn't innovation alone; it's intention, tradition, and humility combined.

Dr Binish Desai

Recycle Man of India

With over two decades in global real estate strategy, I've learned that meaningful transformation starts with small, intentional steps. This manuscript reflects that journey—interweaving environmental resilience, community stewardship, and systems thinking. Its vision aligns with my 'Business to Planet' ethos, where workplace transformation is both sustainable and people-centric. It reminds us that restoring spaces—whether lakes or offices—is ultimately about restoring balance and belonging.

Venkat Neelakantan

Corporate Real Estate Business Leader

As someone who listens deeply to water bodies silenced by neglect, I found *From Breath to Belonging* to be a kindred voice. It is a book that echoes the quiet urgencies I feel when I stand by a dying lake, or watch a sapling take root on a barren bank.

What struck me most was its ability to weave science, mythology, and everyday action into one seamless call—not of alarm, but of awakening. The stories held within this book mirror what I see in the field: that healing the Earth is as much about inner conviction as it is about outer action. This is not a manual. It is a reminder. And sometimes, a whisper is louder than a scream.

Madhuri Subbarao

Environmental Consultant and Founder, Earthitude

With over two decades in cybersecurity, I've learned that true resilience extends beyond digital systems—it must embrace the ecosystems we depend on. *From Breath to Belonging* by Srikanth and Indrajit is a compelling fusion of ancestral insight and ecological foresight. Much like sound security architecture, it urges us to anticipate unseen risks and design with reverence, not just efficiency. A must-read for technologists and leaders committed to building ethically and sustainably.

Balaji Prakash Rao

Chief Digital Product Officer & Deputy CISO, New Jersey

What happens when storytelling meets sustainability, and tradition meets transformation? *From Breath to Belonging* is the answer. Srikanth and Indrajit have written a narrative that feels less like a book and more like a personal rediscovery—of our roots, our responsibility, and our real potential as a civilization. Having spent years championing India's digital transformation and future-ready ecosystems, I often find myself speaking about scale and speed. This book reminded me of a third axis—soul. The stories of Bhoomi, Varaha, and modern-day heroes like Payeng and Arun Krishnamurthy do not preach—they participate in your own inner awakening. It is not just about Scope 1, 2, or 3 emissions. It is about the whispers we've long ignored—from Earth, from epics, from elders. With rare clarity and reverence, the authors bring them back.

Srinivas Rao Mahankali

Ecosystem builder, Ex-CEO T-Hub, Serial Entrepreneur

Every enterprise, whether it serves meals or ideas, must reckon with the question: are we taking more than we give back?

From Breath to Belonging answers this not with prescriptions, but with stories—layered, luminous, and deeply rooted in cultural memory. It reconnects the reader to the soil beneath their success, the breath behind their ambition. For a brand like A2B that cherishes its traditional roots while embracing the modern world, this book resonates deeply. It urges us to look beyond business as usual and recognize that sustainability is not a buzzword—it's Dharma in action.

Ramesh Narayanan

Former CEO, Adyar Ananda Bhavan

From Breath to Belonging is a rare confluence of timeless wisdom and contemporary urgency. As someone engaged in nurturing diverse leadership, I found this book by Srikanth and Indrajit both refreshing and deeply resonant. It doesn't merely preach sustainability; it reawakens our connection to Earth through stories that we thought we had forgotten—of Bhoomi and Varaha, of Yayati and Yu the Great, of Hanuman's leap and Anansi's wisdom. Each chapter unpacks complex ideas like Scope 2 emissions, circular economy, and intergenerational responsibility, but through a human lens that speaks to culture, ethics, and introspection. What struck me most was the gentle yet persistent call to action: that sustainability is not a department or a CSR agenda—it is a mindset.

V Sivakumar

Sr IT Consultant

From Breath to Belonging is a timely and profoundly enriching exploration that beautifully blends ancient mythology with contemporary sustainability principles. Wg Cdr Srikanth and Indrajit Shaw skillfully guide readers on a thoughtful journey, reconnecting humanity with age-old wisdom from around the globe to illuminate our critical role as stewards of the planet.

Each chapter artfully intertwines captivating tales from diverse cultures, whether the philosophical depths of the Vedas, the insightful myths of the Greeks, or the ecological wisdom of indigenous narratives, into a compelling narrative that challenges and inspires.

This book is not just an intellectual exercise, it is a heartfelt call to action, urging introspection, personal responsibility, and collective awakening to the Earth's delicate balance. Its engaging storytelling and poignant reflections make complex sustainability concepts accessible and emotionally resonant.

A must-read for anyone seeking not just knowledge, but a renewed sense of purpose and hope in fostering a harmonious future for our built environment.

Dr Anup Naik

Co-Founder & Director, Urban Frame Pvt Ltd

(A Space Matrix Group Company)

A compelling read for corporate and sustainability leaders, *From Breath to Belonging* reframes environmental responsibility as a timeless leadership value. It blends ancient wisdom with modern challenges, urging us to lead with purpose, balance, and conscience.

Irfan Poonawala

Corporate Real Estate Business Leader

HE FARIH
SO DO

Foreword

Every so often, a book comes along that doesn't just invite you to read—it beckons you to listen. *From Breath to Belonging* is one such book. It doesn't shout or preach; it murmurs softly, much like the Earth itself, inviting you to pause, to reflect, and to reconnect with something far greater than yourself. We live in a world of relentless noise—buzzing phones, clattering cities, the hum of technology. In this cacophony, the whispers of the Earth often go unheard. But if you quiet your mind and open your heart, you'll discover that the Earth is always speaking. It speaks through the rustle of leaves, the sigh of a river, the crackle of a fire, and even the gentle patter of rain. And what it says is both a reminder and a plea: to remember our roots, to restore balance, and to rekindle a sense of wonder and responsibility. This book is more than a narrative; it is an experience, a journey through stories, myths, and realities that illuminate our relationship with the planet. *From Breath to Belonging* weaves a fabric that spans cultures, civilisations, and epochs, drawing threads from ancient wisdom and modern science. Each page is a bridge between the past and the future, a reminder that the lessons of yesterday hold the key to a sustainable tomorrow.

Through its pages, you will travel—from lush forests to arid deserts, from bustling cities to tranquil riversides. You will encounter ancient myths that reveal humanity's timeless bond with nature, and modern insights that shed light on our current challenges. Each story, each whisper, will guide you closer to understanding your place in this intricate web of life. *From Breath to Belonging* is a conversation, not a lecture or a manual. A conversation that meanders gently like a river, carrying you through reflections, realisations, and revelations. It invites you to feel the stories, letting them seep into your being, and emerge with a renewed sense of purpose. The book is a celebration of the resilience of nature, the power of human ingenuity, and the enduring beauty of a planet that continues to give, even as it struggles under the weight of our demands. It reminds us that we are a part of Earth, carrying both the burden of its wounds and the promise of its healing. *From Breath to Belonging* is a journey within yourself—to rediscover wonder, to rekindle hope, and to reimagine the future, enabling you to listen to its joys and sorrows, and respond with compassion and action.

This book is more than words on a page; it is a quiet revolution. It calls you to align your heart with the pulse of the planet, to embrace the responsibility of being a custodian of its future. *From Breath to Belonging* offers something rare and precious: a sense of unity, a shared purpose, and a vision of a future where progress and preservation go hand-in-hand. It is a book for dreamers and doers, for those who seek to understand, and those who yearn to make a difference. Let us listen to the Earth's whispers, not as mere observers, but as active participants in its story. Let us carry its message in our hearts and let it guide our actions. For, in listening, we find the seeds of change; in acting, we nurture them; and in dreaming, we shape a world where every whisper matters.

Let us begin.

Introduction

From Breath to Belonging is an invitation—not just to read, but to listen. To quiet the noise of modern life and tune into the wisdom that has guided civilisations for millennia. This book is not a lament for a dying planet, nor is it a sermon on what we have lost. It is a bridge—a connection between the past and the present, myth and reality, spirituality and science. It weaves together ancient stories and modern struggles, urging us to rediscover the balance that once existed between humanity and nature.

From the dawn of time, cultures across the world have looked to the Earth, not just as a resource, but as a teacher. The Vedas spoke of cosmic order, the Greeks warned of hubris through mythological figures like Prometheus, while the Aboriginal Dreamtime stories depicted a world where humans and nature were inseparable. These tales are not mere folklore; they are blueprints for coexistence. And yet, somewhere along the way, we stopped listening. The Industrial Revolution marked a turning point. Progress became a race, and in our relentless pursuit of growth, we lost sight of the delicate equilibrium that sustained us. We cleared forests, blackened skies, and poisoned rivers—all in the name of progress. And now, as the ice caps melt and wildfires rage, the Earth whispers again.

But this time, it is a cry for help.

Yet, this book is not about despair. It is about remembering. It is about the wisdom of King Prithu, who understood that the Earth's bounty was not infinite and that leadership required stewardship. It is about Lord Ram's teachings to Bharat on the tenets of just rule—principles that extend beyond governance to how we treat the world around us. It is about the cosmic dance of Lord Nataraja, symbolising the eternal cycle of creation, preservation, and destruction—a reminder that every action has a consequence.

As you turn these pages, you will encounter not just stories, but reflections. Questions that demand introspection. Are we, like King Yayati, insatiable in our desires, willing to borrow from future generations to fuel our present? Or can we, like Prithu, restore balance where it has been lost? Can we, like the sages of old, see the Earth as a living entity rather than as a commodity?

From Breath to Belonging is a call to awaken. It urges us to rethink what progress truly means. It challenges us to view sustainability not as a modern trend but as an ancient truth we have forgotten. It asks us to listen—not just with our ears, but with our hearts.

Because the Earth is still whispering. The question is: will we finally listen?

Whispers of the Earth

The Workshop Begins

Namaste, dear friend. I am Lakshmi Gupta, your storyteller and guide.

Let me guide you through this experience.

As you open this book, take a moment to pause and breathe. Inhale deeply, the scent of the Earth after rain, the quiet hum of a world that's alive and speaking; all this can be felt, the only caveat is that you are both, listening and feeling. Close your eyes for a moment and imagine yourself in a lush green garden.

The morning grass feels cool beneath your feet, while your designer yoga mat lies neatly in place. As you assume a yoga posture, deep indigo waves flow into golden mandalas, uniting water and light. Every posture you take rests upon patterns whispering of balance. It feels less like exercise on a mat and more like entering a sacred artwork designed just for you.

As you internalise your environment, your face feels the chill of the moist morning air as it moves gently around your ears. A subconscious feeling stirs, as if whispering to you—yet you cannot make out what it is. Curious, you open your eyes and find others seated beside you in a circle; their faces lit, like yours, with the same curiosity. In the centre of the circle, sit I, Lakshmi Gupta, your storyteller, your guide.

The air is rich with the scent of blooming jasmine and the faint rustle of leaves. I move through the group, softly touching each person's shoulder, speaking to you through this book, with my voice carrying the cadence of ancient tales and modern musings, timeless stories cutting across civilisations. Every story stressing just one thing, 'Maintain Balance.'

As you flip the pages of this book, you aren't just a reader. You are a participant in a workshop. A workshop where the Earth whispers to you through the pages of this book, sharing her joys, her struggles, and her hopes. I am here to ensure that you hear her.

Together let's begin this journey.

The streets are alive and chaotic with the honking of horns, while vendors call out loudly, inviting you to inspect their wares. The faint whiff of *masala chai* brewing at a roadside stall tickles your nostrils.

I will tell you stories of how we arrived at this moment in time with the skies darker than they should be and rivers dirtier than they once were. I live in Janakpuri, a crowded corner of Delhi were humanity hums with relentless energy. My home is a sanctuary, a haven of books, plants, and memories. These are my anchors, my north star, my compass points. They remind me of who I am and why this journey matters.

Together, we'll explore myths and marvels, science and

stories, all stitched together with the golden thread of hope. This book is called *From Breath to Belonging* because that's what it is.

A chance to listen.

A chance to act.

A chance to dream of a world where every whisper matters.

Imagine yourself in a quiet place, free from the rush of the modern world. Now, let me introduce you to someone who has been waiting to speak with you. Listen closely. Can you hear it? A voice, ancient and steady, drawing near.

A deep husky voice fills the room. So resonant it seems to bypass your ears and lodge directly inside your mind.

Do you hear me, my child? I have been here long before fire flickered in the hollow of a fennel stalk, before Prometheus dared to steal fire—the gods' secret—to gift you light.

Something is drawing you towards that voice. You are curious to know whose voice it is.

I am Gaia or Mother Earth, as I am known in Greece, and I have stories to share.

Prometheus

Prometheus, a mythological character, whose name means 'forethought,' is one of the most intriguing figures in Greek mythology. He is best known for his daring and rebellious act of stealing fire from the gods and giving it to humanity. According to the myth, Zeus withheld fire from humans to keep them dependent and less powerful. Prometheus, who had a deep love for humanity and wished to see them thrive and progress, defied Zeus's will. He stole fire from Mount Olympus and brought it to humans hidden inside a hollow fennel stalk.

The gift of fire was transformative.

It allowed humans to cook food, stay warm in the cold, forge tools and develop technology. In essence, fire was the spark that ignited human civilisation. That one act symbolised knowledge, enlightenment, and progress. But this act of defiance came at a great cost to Prometheus. Zeus, enraged by the theft and the empowerment of humans, punished Prometheus severely. He was bound to a rock where an eagle—symbolising Zeus—would eat his liver every day. This punishment continued until he was eventually freed by Hercules.

The fire stolen by Prometheus transformed you. It made you creators, dreamers, and architects of wonder. But it also made you destroyers, forgetful of where the fire was first kindled.

I watched as you harnessed the flame to push back the darkness, forged tools, and built empires. Yet, I have also felt its heat as you burned forests, melted glaciers, and turned the sky into a canvas of smoke.

Just as Prometheus bore the pain of his defiance, I bear the scars of your progress, waiting for the moment you will remember the flame's true purpose—not to consume, but to illuminate; not to harm but to nurture.

My child, your ingenuity is my gift to you.

But every gift comes with a responsibility.

Your choices will decide if I can whisper hope to the generations yet unborn or if my voice will falter into silence. What do you wish to learn from me, my child?

The song of a thriving Earth or the lament of one betrayed?

As Gaia, the provider of everything that you mortals enjoy, let me remind you, modern progress—industrialisation, harnessing of electricity, and improvement in modes of transportation—comes with a significant environmental impact.

Prometheus's act symbolised human ingenuity and the drive to progress, but his story also serves as a cautionary tale about the unforeseen consequences of our actions. His willingness to endure eternal punishment for the benefit of humanity underscores the importance of taking responsibility for one's actions.

Successfully addressing addressing climate change and reducing harmful emissions requires sacrifice and responsible behaviour from individuals, corporations, and governments. Fire symbolises knowledge and enlightenment. Today, awareness and education about environmental issues empower you to make informed decisions and act responsibly.

The Prometheus myth highlights the interconnectedness of all beings.

Prometheus's actions affected not just himself but the entire human race. In your modern world, emissions and environmental impacts are interconnected globally. What happens in one part of the world affects the entire Earth. Imagine yourself as a modern-day Prometheus.

Gaia's voice fades away.

Welcome back! You're here with Lakshmi again! What is your takeaway from Prometheus' story?

You should know that each choice you make, from the energy you use, to the products you buy, has the power to bring positive change or unintended harm. By understanding emissions and taking conscious steps to reduce your carbon footprint, you can harness the metaphorical fire in ways that benefit both humanity and the Earth. Prometheus' story is like a lighthouse, guiding you to use your knowledge and resources wisely, responsibly, and with a sense of care for the world around you.

From Greek mythology, let's turn to our rich Hindu culture, where Bhūmi Devi, your Earth mother speaks to you with more tales.

Bhūmi Devi

Namaskar, I am Bhūmi Devi, your Earth mother.

I have carried you since the first seedling sprouted, since the first rains kissed my soil. You call me by various names—Prithvi, Bhūmi Devi, Gaia, or Earth. And yes, that was me telling you the story of Prometheus as Gaia earlier! I have always loved stories—especially the ones where you and I thrive together.

I nurture, I provide, and I only give.

From the golden grains that grow in your fields to the air that dances in your breath, I give without hesitation, without asking for anything in return. But, my child, even the most generous mother has her limits.

My rivers, once singing with life, now hum a tired tune, their waters heavy with burdens I never placed on them. My soil, once rich and dark, carries an unexpected weight, wrapped in layers of green that do not always nourish. I cradle you still, but oh, how my arms ache—though I smile, always, for you.

Long ago, my story was sung in the verses of the Rigveda.

स नो भूमिं रपदं वेद वि तव्यामृणो धन्वः।
इह देवे वायजन्निह स्थापयेमां सुभगाम्।

It translates to:

> She who knows the Earth and the rain that falls from heaven, She who knows the cows that milk and the path of the winds, O men, let us worship Her in the gods' own habitations, Establishing her in this felicity.

I was once praised for my vastness and my gifts. The sages knew me as the mother who held the cosmos in balance, weaving every tree, every river, every creature into a single, delicate thread. A thread that frays with every careless act.

Nurturing is not just about giving—it is about harmony. I have always worked alongside my kin—Varuna, who moves through my rivers, and Vāyu, whose breath stirs my forests. Together, we shaped a world where life could flourish. You may not always see this balance, but it exists, fragile yet strong, holding everything in place.

There was a time when I, as Prithvi, was one with Dyaus, the sky. We moved as one, Earth and sky, sustaining all beneath our gaze. It was a time of respect and reciprocity.

I have always held you close, cradled you in my arms of soil, water, sky, and wind. Once, there was harmony between Prithvi and Dyaus, Earth and Sky, moving as one. He brought the rains; I nurtured the seeds. His winds carried my whispers; my rivers quenched his thirst. Together, we wove the fabric of life.

But now, the balance is broken. My rivers, once lifelines, are now poisoned. My forests, once endless, stand stripped and silent. The sky, my eternal companion, chokes on smoke and sorrow. I have carried your footprints for millennia, but your steps have grown heavy, pressing deep, taking more.

I have seen this before. Even in the age of the Pandavas, I stood before Arjuna, weary and worn. I told him then, as I tell you now—honour what sustains you. Progress need not be a blade that cuts too deep. Growth need not leave me barren.

I do not rage. I do not punish. I only give. But I am not endless. My voice is in the dry riverbeds, the storm-tossed skies, the lands that no longer bloom. The sky and I still whisper, calling you back to balance.

Can you hear us? Will you listen before the silence takes hold?

Let me repeat Bhūmi Devi's words.

Will you listen before the silence takes hold?

Wasn't it beautiful to hear Bhūmi Devi narrating her

story? Now, I want you to pick up a mirror and view your reflection. You have been told stories about two civilisations that described the imbalance created by humans in the name of progress.

As you view your reflection, think about your contribution to nature's imbalance. But, Bhūmi Devi's stories don't end here. The lessons she offers come wrapped in layers of meaning, each tale a thread in the relationship between humans and Earth.

Varaha

Let us now listen to another beautiful story from *Daśāvatāra*, where Bhūmi Devi, also known as Prithvi, is portrayed as the mother of Hiranyaksha, an *asura* (demon) whose greed and arrogance disrupted the cosmic balance. This asura, in his hubris, kidnapped Prithvi and submerged her in the cosmic ocean, plunging the world into chaos.

What do you think that means?

It symbolises the exploitation and degradation of the Earth, driven by unchecked ambition and disregard for natural limits.

But the story doesn't end in despair.

Vishnu, in his *Varaha avatar*—a mighty boar—descended to rescue Prithvi. He dived into the depths of the cosmic ocean, lifted her on his tusks, and restored her to her rightful place in the cosmos. Balance and order are re-established, a reminder that harmony, once disrupted, demands effort, sacrifice, and sometimes, divine intervention.

This story is more than mythology. It's a metaphor, a reflection of the consequences of disrupting the natural order. It's easy to think technology will save us, but often, it only postpones the inevitable. The Earth will endure, but will we? The story of Varaha and Prithvi reminds us

that restoration isn't just the task of gods. It is humanity's responsibility to act before the balance is irreparably broken.

Ah, do you see where I am leading you? I am not merely telling tales; I am holding up a mirror, hoping you recognise the reflection. Every story whispers the same truth—of balance, of respect, of the delicate dance between taking and giving.

Bhūmi Devi, our Earth, has always nurtured without complaint, yet even she has limits. When taken for granted, she weeps. When plundered, she resists. And yet, she never turns away her children. She only asks for wisdom in return.

The next story is about King Prithu, the one who saw beyond conquest, who understood that a ruler is not a master but a caretaker. When the Earth refused to yield her bounty, he did not force her into submission. He listened. He learned. And in that act of reverence, prosperity was restored.

Listen now, not just with your ears but with your heart. What Prithvi asks of Prithu, she asks of you too. Will you hear her? Will you answer?

Let's begin.

King Prithu

The tale of King Prithu, named after the Earth. It is said that in a time of great strife, the Earth ceased to yield crops, rebelling against the misdeeds of rulers who had exploited her.

Can you imagine an Earth that refuses to nurture?

The very lifeline we depend on, withdrawing her gifts as a form of protest? King Prithu, a righteous sovereign, took it upon himself to restore Earth's fertility. He chased her, determined to correct the imbalance, but not through force—through governance, ethics, and care.

Prithvi, recognising his integrity, submitted to his rule and promised to be fruitful under his just reign.

This tale, ancient yet ever-relevant, teaches us that the prosperity of the Earth depends on those who lead, on their commitment to equity, and ethical stewardship.

Prithvi, in these stories, is not just a passive entity. While divine interventions like Varaha's rescue are awe-inspiring, the story of King Prithu reminds us that mortal efforts, too, have the power to restore balance, provided they are driven by justice and reverence for the natural world.

What I am saying is simple. We have one Earth. There is nowhere else to go if we degrade it beyond a point. Earth has supported our exploitation of her gifts throughout human history. She has been more than patient but tell-tale signs are visible through floods, drought, famine, heatwaves, fires, the melting of ice caps, frequent cyclones, just as we have been seeing in our stories. The root cause of all these calamities points to one thing: Imbalance between man's progress versus nature's balance.

I have seen graffiti in numerous places in Delhi where black paint sprayed across walls said, 'Where have all the sparrows gone?' It was the cry of environmental activists who aren't taken seriously at all and lo behold, we have a number of birds that have become rare in a dense urban environment.

If you look around my home and observe, you will notice I have House Sparrows, the Indian Peafowl, the Common Hoopoe, Barn Owls, Purple Sunbirds, the Common Kingfisher, the Indian Roller, the Black Drogo, the Asian

Koel, and Coppersmith Barbets, within my garden.

Once ubiquitous in cities, all these birds have become rare due to the loss of their nesting sites due to increased concretisation of cities, reduced availability of food, reduction in green spaces, human interference, overuse of pesticides and spraying in urban areas, loss of flowering in plants, and pollution.

I don't know how it happened, but due to the way in which I actually live and the lifestyle I consciously maintain, these birds appeared gradually in my garden over a period of eight months and today this is no less than a rich biodiversity reserve.

Lord Nataraja

Let us now turn to the eternal dance of Lord Nataraja. He is a symbol of balance, one that resonates with the whispers of the Earth itself.

Lord Nataraja, the cosmic dancer, embodies the cycle of creation, preservation, and destruction—the rhythm that sustains the universe. His dance, the *Ananda Tandava,* is a visual hymn to this balance, a reminder that life flourishes in harmony but falters in chaos. This balance is not confined to celestial spheres. It echoes in our lives and extends to the delicate equilibrium of the Earth.

Within us, this balance is achieved through *Sāttvic* living—a life that is pure, simple, and in tune with nature.

When we adopt a Sāttvic lifestyle, we attune ourselves to the rhythm of Lord Nataraja's dance. Our food is wholesome, nourishing the body without excess; our thoughts are peaceful, elevating the mind; and our actions are selfless, benefiting not just ourselves but the larger world. This alignment creates harmony, both within and without. And when we live in harmony, the Earth responds—abundantly, gently, with whispers of approval carried by the wind and the rivers.

Yet, when we deviate from this path, embracing *Rājasic* or *Tāmasic* living, the balance is disrupted.

Rajas, driven by restless desire and unchecked ambition, pushes us to take more than we need, to tread heavily on resources, and to chase endlessly after illusions of progress.

Tamas, with its inertia and ignorance, blinds us to the consequences of our actions, leaving us disconnected and apathetic.

The disruptions we see in the natural world today—extreme climates, dwindling resources, and fractured ecosystems—are but reflections of this imbalance within us. The Earth's whisper has turned into a lament, much like Prithvi's plea to Arjuna, urging us to act with awareness before the imbalance becomes irreversible.

To restore this harmony, we must return to Sāttvic living. This is not merely a personal choice but a collective responsibility. When we simplify our needs, consume with mindfulness, and act with awareness, we realign ourselves with the natural world. In doing so, we become part of the Tandava, moving in sync with the cosmic rhythms.

This is not a call to retreat from progress but to redefine it. Sustainability is not just about solar panels or recycling programmes—it begins within. It is an internal state of balance, a Sāttvic harmony that ripples out into the world. Lord Nataraja's dance teaches us this truth: sustainability is not an external obligation but an eternal *Dharma*, a duty woven into the fabric of our existence.

Imagine a world where humanity moves in step with Nataraja's cosmic dance. Where each decision—whether to build, consume, or discard—is made in awareness of its impact on the Earth. When we find that balance within, the Earth responds, not with chaos but with calm, not with scarcity but with abundance.

This is the wisdom of Lord Nataraja. His dance is our guide, and his rhythm, the whisper of the Earth, calls us to step in tune. Will you join the dance?

We search for answers in the hustle of today, but sometimes, the most profound wisdom is found in the whispers of the past.

Our scriptures, are not merely stories from a forgotten time. They are seeds planted for a future we are only beginning to understand. The answers you seek, the path to sustainability—they have always been there, just that we don't see or hear them well enough.

Ram and Bharat

Take the moment between Lord Ram and his beloved brother, Bharat. You might remember the image of Bharat carrying Ram's *paduka*, returning to Ayodhya with them as a symbol of his devotion. But what lies deeper, in the words exchanged between them, are principles that can guide anyone—whether a king or a commoner—towards a life of balance, compassion, and sustainability. Their conversation is not a

fleeting exchange; it is a timeless dialogue, as relevant today as it was then. Ram's words to Bharat were like rivers carving their path through the landscape of his heart. Those ten principles—they are not just for rulers, my friend. They are for all of us, in every corner of the world, seeking a way to live with wisdom and harmony.

And so, it goes like this between them.

Bharat: *Brother, Ayodhya yearns for your return. The kingdom needs you. I am not worthy to rule in your place. Please, return and guide our people.*

Ram: *Dear Bharat, the path of Dharma is not always the easiest, but it is the righteous one. It is my duty to fulfil our father's command. You are more than capable of ruling Ayodhya with wisdom and compassion. However, if you must lead, there are ten principles you should embody. These principles will not only ensure a prosperous kingdom, but also a sustainable future for our people and the Earth.*

- *A king's foremost duty is to serve his people with love and compassion. Their welfare should be your top priority.*
- *Always be just and fair in your decisions. Justice is the bedrock of a peaceful and stable society.*
- *A king must protect the forests, rivers, and all natural resources, for they are the lifeblood of the kingdom.*
- *Ensure that your economy is stable, with prosperity reaching even the poorest of your subjects.*
- *Foster wisdom and knowledge in your kingdom. Education is the light that dispels the darkness of ignorance.*
- *Uphold the law with integrity and ensure that peace and order are maintained in your kingdom.*
- *Lead by example, Bharat. Your actions should reflect the virtues you wish to see in your subjects.*
- *Honour the traditions and culture of our land, for they are the soul of our people.*
- *Exercise restraint and humility in your rule. Do not be swayed by power or pride.*
- *Always think ahead, Bharat. A wise king plans for the future and leaves behind a legacy of prosperity and peace.*

Bharat: *I will, dear brother. I shall lead Ayodhya with the wisdom you have bestowed upon me, always keeping the well-being of our people and our land at the forefront of my rule.*

Let's step into Bharat's shoes, for a moment, and feel the weight of his responsibility. Imagine that you are not just responsible only for your loved ones but for the very soul of the Earth itself.

Experience the world through the eyes of a guardian—one whose duty it is to safeguard the world, its people, and its essence. Don't you feel your heart swell with the responsibility, feeling the connection to something larger than yourself?

Hear Lord Ram's words not as a distant myth, but as a living truth.

These ten teachings are not abstract; they are not mere ideals. They are the breath of life. They are a guide for every step you take, every action you perform, and every decision you face.

Lord Ram's words are for everyone who has the privilege of walking this Earth. A blueprint, for humans to live in harmony, with nature, with each other and with themselves.

Lord Ram's ten principles were the very foundation of the Sustainable Development Goals.

They weren't merely policies or targets, they were ancient truths, revived in modern language. We were only just beginning to listen. We were not merely stewards of the Earth; we were part of it. We were woven into the fabric of the natural world in a way that went far beyond our daily lives, our personal achievements, and even our survival. We were a living extension of the Earth's story, just as our ancestors had been before us.

Look at the world through this lens, and you'll begin to see the interconnectedness of all things. Look closely at the ten principles of Lord Ram. The Sustainable Development Goals are not something new; they are the same teachings, repackaged for the modern world. They are the whispers of the Earth, guiding us toward a future where humanity lives in balance with the planet.

This is the whisper of the Earth, the quiet call to all of us who are willing to listen, to remember who we are and where we come from. We are not separate from the Earth; we are part of her. And as we protect her, we protect ourselves. As we honour her, we honour the very essence of life itself. It isn't just a duty to protect the Earth—it is a sacred calling.

Smoky Beginnings

Turning Gears and Blackened Skies

Let us call this section 'Sparks of Change.' Before we step into our time machine, any guesses on where we will go next?

Let me give you the answer. Next, we'll be travelling back to the late 1700s, right to the heart of the Industrial Revolution. Imagine that you are there, walking through a city.

Close your eyes and feel the energy of this transformative moment.

As you step out into the world, you're greeted by the sight of factories, their towering chimneys bellowing thick clouds of black smoke into the sky. The air smells heavy—like burning coal and sweat, a far cry from the fresh winds that once kissed the Earth.

This was a time of enormous change. People poured into the cities from rural areas, drawn by the promise of jobs in the factories. But was it progress? Was human life better?

I would say, no, not necessarily better, but more affluent. We see human endeavour shaping machines that make life easier, more efficient. And yes, the machines worked miracles in transforming industries. But for every action, there is a consequence—an intended result and an unintended one.

The intended result of these factories was clear: a world transformed by industry. But the unintended consequence was just as powerful—the smoke, the pollution, the thick clouds hanging in the sky.

In the context of sustainability, these factories of the past serve as a powerful metaphor. The smoke from those towering chimneys is the earliest form of what we call Scope 1 emissions today.

Let's break it down. Scope 1 emissions are the direct emissions from sources that you own or control. They come straight from the source, no middleman.

Think of it this way, if your business runs a diesel generator to power your premises, the emissions from that generator are Scope 1 emissions. Direct, undeniable, and under your control.

This is where the whispers of the Earth begin to speak to us. You see, the Earth has always been telling us something. From the rise of industrialisation in the 1700s to our present-day emissions, the Earth's whispers have been consistent.

Learning from the Bhagavad Gita

In the Bhagavad Gita, Lord Krishna tells Arjuna:

> *As a fire, when covered by smoke, is obscured, so is the living entity, when covered by material nature, is obscured.*

In the same way, the Earth's clear voice has been clouded by the smoke of human actions. The message has always been there, but we haven't been able to hear it through the noise we've created.

The story of Scope 1 emissions isn't just about pollution—it's about awareness. It's about listening to the Earth's quiet call and understanding the direct impact of our actions. We cannot escape the responsibility that lies at the source. Just as those towering chimneys poured out smoke into the sky, every decision we make today sends a ripple through the Earth. We can no longer pretend that these ripples don't matter.

This is where the whispers of the Earth call to us, asking us to listen and to act. The unintended consequences of our industrialisation—the smoke, the carbon, the emissions—provide us with an opportunity to listen, to realise the balance we must restore.

See the entire picture—the echoes of industrialisation, the rise of emissions, the lessons from the past, and the call to act now. The whispers of the Earth are not just ancient stories—they are alive in every choice we make, in every action we take.

Learning from Srimad Bhagavatham

The Industrial Revolution brought incredible progress, but it also marked the beginning of significant environmental impact. Recall where it is said in the *Srimad Bhagavatham*, that the Earth and its elements must be treated with respect and balance.

स वै पुंसां परो धर्मो यतो भक्तिरधोक्षजे ।
अहैतुक्यप्रतिहता यया यात्मा सुप्रसीदति ॥

(Srimad Bhagavatham 1.2.6)

The supreme occupation for all humanity is that by which men can attain to loving devotion unto the transcendent Lord. Such devotional service must be unmotivated and uninterrupted to completely satisfy the self.

This verse highlights the supreme duty of humanity and the importance of unmotivated and uninterrupted devotional service, implying a need for balance and respect towards nature.

Industrial Revolution

The Industrial Revolution was a time when that balance began to tilt.

Before the Industrial Revolution, life was deeply rooted in agriculture, where communities thrived in harmony with the land. People lived in villages, working the soil, growing crops, and raising livestock. The rhythm of life was in tune with the seasons, and nature provided sustenance. There was a delicate balance between human activity and the environment—an unspoken respect for the

land that sustained all life. The landscape was dotted with small farms, and the air was fresher, the Earth more fertile. But then, the Industrial Revolution arrived like a sudden storm. In the late eighteenth century, the introduction of mechanised manufacturing shifted the entire course of human civilisation. The steam engine, perfected by James Watt, became the heartbeat of this transformation.

Machinery replaced manual labour, and productivity surged. As factories grew, they became the new centres of human activity. But the rapid mechanisation required one thing above all else—coal. Coal became the fuel that powered the new engines of progress, and with it came a dark cloud that would forever alter the environment: the smoke from factory chimneys. This smoke—thick, choking, and pervasive—marked the dawn of a new era. It was the first tangible manifestation of Scope 1 emissions, direct emissions released from sources controlled by industries. Just as these factories burned coal to power the engines that drove innovation, they also burned the balance between humanity and nature. In a world where humans had once lived in synchrony with the seasons, the Industrial Revolution now saw them tearing apart the very air they breathed, the water they drank, and the soil they relied on. The echoes of this transformation reached far beyond the factories of Europe.

Iblis, the fallen angel

The story of Iblis, the fallen angel, offers a striking parallel to this period of industrialisation. Iblis, in his arrogance, refused to bow to Adam, believing that his creation from smokeless fire made him superior to the humble clay from which Adam was made. His refusal to acknowledge the divine balance led to his downfall. In a similar vein, the Industrial Revolution represented humanity's refusal to bow to the natural world—an arrogance that led to the disruption of the environment. Just as Iblis' pride blinded him to divine wisdom, the industrial age blinded humanity to the need for balance with nature. But unlike Iblis, we have the chance to learn from our mistakes. We can acknowledge the need for balance, for respect, and for humility. Just as Iblis fell from grace for his defiance, we must learn that progress, if not tempered with sustainability, can lead to irreversible damage.

The black smoke of the past still lingers in our skies, and the scars of industrialisation remain on Earth. But the lessons of Scope 1 emissions—of unchecked arrogance and the inevitable fallout of ignoring nature's balance—are lessons we must heed. Our progress must not come at the expense of our planet. The whispers of the Earth, though soft and often ignored, call out to us, urging us to reconsider the path we are on.

The journey ahead calls for humility. It calls for recognition that progress, when coupled with respect for the Earth, can lead to a future where both humanity and nature thrive. Let us not be blinded by arrogance as Iblis was, but instead, let us listen to the whispers of the Earth and restore the balance that has been lost.

Transformation of Society

It is not the strongest of the species that survive, nor the most intelligent, but the one most responsive to change.

-Charles Darwin

The world was changing, and so too were the lives of those who inhabited it. The advent of factories, with their rhythmic clanging and constant churn, brought with it a promise of better days—of jobs, of prosperity, of dreams realised. For centuries, humanity had cultivated land, lived with the seasons, and built close-knit communities. But with the rise of industry, drawn by the lure of progress, people left their fields and forests behind. They journeyed to the cities in search of work. Soon, migration turned into a great tide that swept away the old ways. In droves, they traded the quiet rhythm of rural life for the electric pulse of the industrial machine. Cities like Manchester and Birmingham swelled in size, their skylines punctuated by the towering chimneys of factories that seemed to promise an endless supply of opportunities.

But what did these smokestacks conceal?

A darker truth.

The very same factories that fuelled economic growth and offered a glimmer of hope to the downtrodden, also cast a shadow over the land. The air, once fragrant with the scent of Earth and green fields, grew thick and heavy with soot, the kind that clung to clothes, to homes, and lodged inside the very lungs of people. The once clear skies, under which families had gathered and children had played, were now painted a dull grey, the sun was a distant memory, shrouded behind a perpetual veil of smoke.

This was the price of progress—the sacrifice that accompanied the rise of modernity. The city buzzed with energy, but it was a harsh energy, one that came at the cost of the world around it. The streets, now crowded with carts, workers, and hurried footsteps, no longer carried the soft hum of nature. Instead, they reverberated with the harsh clang of industry, and beneath the noise, a new, less visible but far more insidious presence began to take hold: pollution. Pollution was not just something you could see. It was something you could feel—the heaviness in your chest, the burning in your throat, the cough that never seemed to go away.

Carbon dioxide, the silent driver of climate change, slowly gathered in the atmosphere, a steady, unseen accumulation that began to change the world's weather patterns. The particulate matter, though invisible to the eye, was far from harmless. These tiny particles—like seeds of illness—penetrated deep into the lungs of those who lived in these cities, bringing with them diseases that would mark a generation. This environmental cost—the smog that blurred the lines of sky and Earth—was not just a byproduct of industry; it was the price paid by every person who sought a better life in the city. And yet, even as the cities grew and the factories continued to churn, the social fabric of life began to shift in unexpected ways.

The Industrial Revolution, for all its destruction, also carried with it the seeds of positive change. The standard of living began to rise. Families that had once scraped by on meagre incomes now had access to goods and services that had once been the realm of the wealthy. The middle class expanded, its appetite for consumption growing, and with it, a new social mobility emerged. People, once bound by the constraints of rural poverty, now saw the possibility of a different future. But this new world was not without its cost. The slow erosion of cultural ties was palpable. The rural way of life, where the rhythm of the seasons governed everything, was replaced by a faster-paced, capitalistic culture that prized individual success above all else. It was a society built on the idea that progress was synonymous with consumption, that happiness was measured by the goods you owned, the work you did, and the status you attained.

In the face of such a transformation, it is the stories of the past that offer the greatest wisdom.

The Tale of King Yayati

Lesson in Balance and Responsibility

Now we're travelling to the era of the Mahabharata, where the lessons of balance, ambition, and consequences are deeply embedded in stories even today.

Yayati was a ruler of immense power and prosperity.

He was a just king, beloved to his people, ruling a kingdom that flourished under his guidance. His reign was marked by tremendous wealth and economic prosperity, much like the Industrial Revolution's promise of affluence and power. Yet, just as the factories of the 1700s began to pollute the skies, Yayati's insatiable desire for more—more power,

more wealth, more pleasures—led him down a dangerous path. Yayati was married to Devayani, the daughter of Sage Shukracharya. However, his heart wandered, and he fell in love with another woman, Sharmishtha. This action, born out of desire, led to his downfall. For breaking his promise to Devayani, Yayati was cursed by Shukracharya to age prematurely, a punishment for his uncontrolled pursuit of pleasure and material wealth. But here's where the story takes an interesting turn.

King Yayati, unwilling to accept his sudden old age, pleaded with his sons to exchange their youth for his old age, so he could continue enjoying the pleasures of the world. His youngest son, Puru, selflessly agreed. Yayati lived for many more years in the full bloom of youth, indulging in every possible pleasure, much like the industrialists of the eighteenth century who exploited natural resources for prosperity without regard for long-term consequences.

Yet, after years of indulgence, Yayati realised that his desires were like a fire—they only grew stronger the more he fed them. He came to understand that true contentment cannot come from unchecked pursuit of pleasure or wealth, but from balance and self-restraint.

Realising his mistake, he returned Puru's youth, abdicated the throne, and sought peace in the forest through meditation and a simple life. His story teaches us of the consequences of losing balance between desire and responsibility.

Does this not remind you of the Industrial Revolution? The rapid industrial growth, driven by coal and steam engines, was like King Yayati's relentless pursuit of more—more machines, more factories, more wealth. But just as Yayati's unchecked desire brought unintended consequences, the

Industrial Revolution brought environmental degradation, pollution, and the imbalance we still grapple with today.

The curse of ageing prematurely was a metaphor for the degradation of nature. While Yayati's personal imbalance affected his family, the imbalance created by industrialisation impacted the Earth. The black smoke rising from factories, the pollution in rivers, and the warming of the planet—these are the premature ageing of our world. They are the consequences of our actions, just as Yayati's ageing was the result of his choices.

Desire without restraint leads to imbalance and decay.

Yayati's realisation came too late to reverse the ageing caused by the curse, but he did find peace in recognising the importance of balance.

For us, it's not too late. We can still correct the imbalance between industrial progress and environmental stewardship. Just as Yayati found peace by renouncing his insatiable desire, we can find harmony by rethinking our relationship with nature, by embracing sustainability and respecting the limits of our planet.

Mahatma Gandhi, in his wisdom, reminds us of our duty to the Earth:

We have a moral responsibility to take care of the environment and ensure that future generations have access to clean air and water.

It is not just a call to action—it is a moral imperative. As we look toward the future, it is our responsibility to address the environmental challenges that have been set in motion by industrialisation. We must learn to balance the scales—embracing progress while safeguarding the Earth for those who will come after us.

The transformation of society during the Industrial Revolution was immense, but its legacy is not just one of factories and cities—it is the story of how we, as humans, respond to change.

It is a story of loss and gain, of progress and sacrifice, of connection and isolation. And it is a story that continues today. The choices we make now, in the face of climate change and environmental degradation, will define the world we leave behind. Like King Yayati, we must learn to balance our desires with our responsibilities, to find a path that leads not just to prosperity, but to sustainability, to harmony with the Earth, and to a future where both people and the planet can thrive.

I know that at first, this subject might seem dry, perhaps even distant, but I am certain that once the seeds of sustainability are planted, the desire to live in balance would become second nature—something addictive. This, I believe, is the true essence of the workshop itself, and I am determined to guide you towards that.

Gandhi's call for moral responsibility is more relevant today. We must acknowledge the lessons of the past—both the triumphs and the mistakes—and take proactive steps to reduce emissions. In doing so, we honour our responsibility to the planet and to future generations, ensuring that the progress we pursue does not compromise the health of the Earth.

That was a lot to digest, wasn't it? Alright, let's shake off the cobwebs from our minds after that deep dive into history. Imagine for a moment that you are part of the Industrial Revolution—a time of great change, but also great consequence. Let's cleanse our thoughts and spirits a bit, shall we?

Now, close your eyes and imagine this: You are sitting by the Ganges, the gentle flow of the river beside you, the soft rustling of leaves in the air. You sip on a steaming cup of warm *masala chai*, the rich aroma wrapping around you like a comforting embrace.

The mention of masala chai, with its comforting warmth and rich spices, seemed to bring the words to life. The imagery was so vivid that, as you read, you could almost feel the warmth of the cup in your hands, the gentle steam rising from the chai, mingling with the fresh air by the Ganges. You could almost taste the spices—cardamom, cinnamon, and ginger—on your tongue. The scene she painted was so real, so inviting, that it enticed you, the reader, to pause for a moment. As if on cue, you found yourself setting the book aside and heading to the kitchen, drawn to the idea of brewing a cup of masala chai, just as Lakshmi had suggested.

What I am going to do now is to share some ancient wisdom from our scriptures. But don't worry, I promise there won't be a pop quiz at the end. But you will be compelled to think about why we don't read more or someone doesn't tell you more about this topic.

Our ancestors had a strong understanding of balance and respect for nature. Let's dive into the teachings from our sacred texts that provide timeless guidance on how to live harmoniously with our environment. Mind you, we are still in the 1700s where at that time all humans did was light a fire to perform any task. These fires were all Scope 1 emissions. There was nothing beyond that.

With that said, let's see what the Bhagavad Gita, the

Srimad Bhagavatam, the *Shiva Purana*, and reflections on *Pralayam* (the concept of cosmic dissolution) say, weaving ancient wisdom into the modern context of Scope 1 emissions.

The Bhagavad Gita offers amazing insights into the balance between progress and environmental stewardship. In Chapter 3, Verse 16, Lord Krishna says:

evaṁ pravartitaṁ chakraṁ nānuvartayatīha yaḥ
aghāyur indriyārāmo moghaṁ pārtha sa jīvati
One who does not follow the cycle of sacrifice established in the Vedas lives in vain, full of sin and indulging in sensual gratification.

This verse emphasises the importance of following the natural cycle and living a life of balance and responsibility. The Industrial Revolution, though essential, disrupted this cycle, prioritising rapid industrial growth over the well-being of the environment. The uncontrolled emissions from factories were a form of disregard for the natural order, leading to significant environmental harm.

The *Srimad Bhagavatam* also highlights the consequences of disturbing the natural balance.
In Canto 1, Chapter 17, Verse 10, it is mentioned:

dharmena hīnāḥ paśubhiḥ samānāḥ

Men who have lost their virtue are no better than animals.

This verse underscores the importance of Dharma, or righteous living, which includes respect for nature. The Scope 1 emissions from industrial activities represent a departure from Dharma, resulting in pollution and environmental degradation. By ignoring the natural laws and focusing solely on economic gains, humanity faced the consequences of polluted air and deteriorating health.

Shiva Purana

The *Shiva Purana*, another significant scripture, speaks of the consequences of imbalance and neglecting nature.

In the *Rudra Samhita, Yuddha Khanda,* Chapter 4, Verse 12, it is stated:

prakṛtiṁ viddhi me parām

Understand that nature is my supreme manifestation.

This verse conveys that nature is a divine manifestation and must be treated with reverence. The unchecked emissions during the Industrial Revolution were a direct violation of this principle, leading to environmental damage. The acknowledgment of Scope 1 emissions today reflects an attempt to rectify this historical oversight and align with the teachings of respecting and preserving nature.

The concept of Pralayam, or cosmic dissolution, described in Hindu mythology, symbolises the destruction that follows when the balance of nature is disrupted. In the context of Scope 1 emissions, this serves as a stark reminder of the potential consequences of environmental neglect.

Vishnu Purana

As described in the *Vishnu Purana:*

kalpante pralayo 'pi nityaḥ

At the end of the cycle, dissolution is inevitable.

The excessive emissions from industrial activities can be seen as a microcosm of this larger cosmic cycle, where disregard for the environment leads to its degradation.

Linking Scriptures to Today's Scope 1 Emissions

The teachings from the Bhagavad Gita, the *Srimad Bhagavatam*, the *Shiva Purana*, and the concept of Pralayam, collectively emphasise the importance of maintaining balance and respecting nature. Scope 1 emissions, being direct pollutants from industrial activities, highlight the consequences of deviating from these ancient principles. The current efforts to monitor and reduce these emissions are steps toward restoring the balance and aligning with the moral codes outlined in our scriptures.

Now imagine, if you will, Earth as a woman—soft but tough, wise, and ancient—whispering her tale into your ears.

I take you back to a time of smoke and steel,
Where factories rose and progress felt real.
Men and women, in search of light,
Fled the fields to cities bright.
But the price of progress, oh, it was steep—
The air turned dark, and the rivers weep.
In the rise of growth, the balance was lost,
And pollution paid the heavy cost.
Gandhi once called, his voice so clear,
'Protect the Earth, for those we hold dear.'
Now hear my plea, as I come near—
Let's move forward, but with love and care.

Watt's Up

Era of Power Surge

We will call this section, The Power Surge to Clean.

Picture a world on the cusp of transformation and you are standing at the edge of the nineteenth century as time chugs on. The streets glimmer faintly under gas lamps, and households rely on candles to fend off the night. Electricity, in that era, wasn't just a technical curiosity–it was a promise, whispered by inventors who dared to reimagine life itself.

When Thomas Edison introduced the incandescent light bulb, it wasn't just a marvel of engineering. It was a moment of awe—a quiet declaration that darkness could be conquered and life illuminated. But for all its brilliance, the story of electricity has always been one of duality. Its arrival sparked industry, ignited imagination, and illuminated the lives of billions. Yet, hidden in its glow was another truth—a cost that wouldn't reveal itself until much later, when the air grew warmer, the seas rose higher, and the planet began to whisper of its warming.

Most of us rarely pause to think about where our electricity comes from. We flick a switch and the world brightens. We plug in our devices and they hum to life. Its seamless, invisible—an act so mundane we forget its implications. But every kilowatt hour of electricity consumed carries a story.

Maybe it begins at a coal-fired plant, where fossil fuels are burned to generate power, releasing carbon dioxide into the atmosphere. Or perhaps it starts with a hydroelectric dam, harnessing rivers to turn turbines. These stories form the invisible ledger of what we call Scope 2 emissions. Just another term meaning indirect greenhouse gases tied to the electricity, the heating, or the cooling that we consume. Unlike the smog that once darkened cities, Scope 2 emissions are harder to grasp. They don't rise from chimneys in our neighbourhoods. Instead, they ripple out from distant power plants, linking our everyday actions to global environmental impact. The story of electricity is still being written. It is about rewriting our relationship with power itself.

As we illuminate this new path, the question isn't just technical: How do we decarbonise? It is also moral: How do we reconcile progress with responsibility? And in answering that, perhaps we can create a future in which every light that is switched on brings not just comfort but also hope.

This is where conscience meets practicality.

Transitioning to renewable energy isn't just a corporate or governmental responsibility. It is a collective effort. Whether it is a household installing solar panels, a company investing in wind power, or a community advocating for clean energy policies, the shift begins with recognising that electricity is not a limitless gift, it is a resource that demands stewardship.

But what happens when a civilisation sees beyond mere

consumption? When it perceives energy not just as a tool of power but as a sacred trust? Some of the most enduring legacies come from leaders who understood this balance, who recognised that true strength lay not in unchecked dominion over resources but in the wisdom to sustain them.

Let's journey into the annals of three civilisations, where hidden legacies and lessons await. Through their tales, we might uncover not only the wisdom of the past but also a mirror reflecting the unseen responsibilities of our present.

King Bharata's Legacy: Beyond the Visible

In ancient Bharat, King Bharata, the eponymous ruler of the land, was revered not just for his might, but for his understanding of *Dharma*—the eternal law of harmony.

From the towering peaks of the Himalayas to the whispering shores of the southern oceans, his kingdom thrived under his reign. Yet, despite the peace and prosperity that enveloped his realm, Bharata was a man given to introspection. Each morning, he would rise with the sun, its first rays cascading over his palace and casting long shadows on the land he governed.

He would climb to the highest point of his citadel and gaze at his kingdom. He saw the rivers carving their way through fertile plains, the forests alive with the hum of life, and villages where laughter mixed with the music of work.

Yet, beyond this beauty lay an unshakeable question:

What makes a ruler great?

One evening, Bharata held court under a banyan tree whose roots seemed to echo the wisdom of ages. Among the many who came to seek his judgement and blessings was an elderly frail villager.

In his weathered hands, he carried a single flower.

The king accepted the humble offering with curiosity. 'What is the story behind this flower, old one?' Bharata asked gently.

The man's voice trembled, yet his words carried an unyielding truth. 'My lord, this flower blooms in a field that flourishes because of your wisdom. Though your feet have never graced our soil, your decisions have nurtured our lives. The irrigation canals you decreed brought water to our parched lands. The granaries you ordered built ensured we never starved in the harshest of seasons. This flower is but a token of our gratitude.'

Bharata sat in silence. He realised that his power was not confined to his palace walls or to the reach of his armies. It extended far beyond—like the rivers and winds that carried life to places unseen. His whispered decisions, made in the quiet sanctity of his court, shaped lives in distant hamlets he would never visit.

That night, Bharata wrote in his journal:

The unseen legacy of a king is far more enduring than the visible.
The measure of a ruler lies not in the grandeur of their throne
but in the quiet flourishing of lives they may never know.

Fast forward thousands of years, and the whispers of King Bharata's wisdom echo in the challenges of our modern world. The decisions we make today, especially in how we consume energy, bear a striking resemblance to Bharata's invisible legacy.

Scope 2 emissions—those indirect greenhouse gases resulting from purchased electricity, heating, or cooling—are the unseen consequences of our everyday actions. When we flick on a lightbulb, charge our devices, or power our homes, we often think only of the immediate convenience. Rarely do we consider the journey of energy—the coal burned in a distant plant, the natural gas consumed, or even the hydropower generated far away. These actions leave an imprint, a ripple in the environment, though they seem disconnected from us. Much like Bharata realised that his decrees affected villages beyond his sight, we must recognise that the energy choices we make in our homes and workplaces affect the broader world. A light glowing in one room might lead to smoke rising from a stack thousands of miles away, and yet the responsibility remains ours.

Imagine standing in Bharata's shoes, atop the citadel of today's energy landscape. The rivers are now energy grids; the villages are global ecosystems. The decisions we make—whether to invest in renewable energy, improve efficiency, or reduce consumption—carry the weight of unseen impacts.

Just as Bharata used foresight to nurture unseen villages, we must adopt a similar vision for Scope 2 emissions. The switch to renewable energy sources like solar, wind, or hydro is not just an investment in cleaner air; it's a commitment to a future where the unseen harm is minimised. Every kilowatt-hour saved or 'greened' is a step toward ensuring that distant 'villages'—the ecosystems, communities, and generations yet unborn—can thrive.

King Bharata's story teaches us the power of humility and responsibility. He ruled with the knowledge that his decisions would outlive his reign, shaping lives he might never know. In the same vein, our energy choices today are not merely transactions—they are legacies in the making.

Reducing Scope 2 emissions is not an abstract corporate

goal; it is deeply personal and human. In corporate terms, it is akin to an individual's key performance indicator. It means lighting up our lives with the same care Bharata showed his people—knowing that even if we don't see the coal plants or the emissions, they are as real as the unseen villagers who blessed Bharata's reign. When you next flip a switch or plug in a device, remember Bharata and his unseen legacy. For in the hum of the electricity that powers our world lies a responsibility—one that calls us to be rulers of our own unseen kingdoms, caring for what lies beyond our immediate sight but remains forever in our charge.

Samudra Manthan–Churning of the Ocean

Link to a Modern Pulse

There was a time when the heavens held their breath, and the Earth whispered tales of balance and struggle. In the great cosmic ocean, a momentous endeavour unfolded—a collaboration born not of trust but of necessity. The gods and demons, sworn adversaries, set aside their enmity for a single, extraordinary goal: the nectar of immortality, the *Amrit*.

Yet, the story was never just about the nectar. It was about the process, the cost, and the truths that emerged from the depths.

Mount Mandara was chosen as the churning rod, unyielding and steadfast. Vasuki, the great serpent, coiled around it as the rope. As the gods and demons pulled and released in rhythm, the ocean began to churn. Waves rose, frothing with anticipation and treasures of untold beauty surfaced: celestial gems, divine animals, and secrets of the universe.

For a fleeting moment, the gods were spellbound, their

hearts beating with the promise of prosperity. A dense, noxious poison, *Halahala*, erupted from the waters. The skies darkened, and the very air seemed to shudder with its deadly presence. The gods faltered, their unity unravelling. The demons, too, recoiled, recognising the peril that threatened not just their enemies but existence itself.

In the face of this calamity, Lord Shiva stepped forward. His presence was serene, almost otherworldly. He did not speak of blame or fear. He simply acted. Taking the Halahala in his hands, he drank it, containing its destruction within himself. His throat turned blue, a mark of his sacrifice. It was a moment of profound humanity from a divine being—a reminder that true power lies in bearing the burdens others cannot.

This ancient tale is not confined to the realm of mythology. It lives within us, unfolding in the choices we make every day. Today, our churning takes a different form. We mine the Earth, drill deep into its crust, and harness the energy that powers our modern lives. The treasures we extract—coal, oil, gas—are the gems of our age, lighting our cities and fuelling our dreams. But with every treasure comes its shadow. The Halahala of our time rises as invisible carbon emissions, poisoning the air we breathe and the world we share. It creeps into the atmosphere, disrupting climates, melting glaciers, and unsettling the balance of life. And just as the gods and demons once stood paralysed, we, too, find ourselves at a crossroads.

The story of *Samudra Manthan* is not just a story of gods and demons. It is a story of humanity—of how we respond when faced with the dual outcomes of our actions. The treasures of progress come at a price, and like Shiva, we must find the courage to bear the cost. Each of us holds a spark of that divinity within.

Every small step—a switch to solar panels, a decision to use energy wisely, a commitment to advocate for renewable resources—is an echo of Shiva's act. It is the quiet heroism of taking responsibility, even when the challenge seems insurmountable. Imagine for a moment that you are a part of this great churning. The choices you make—the light you turn off, the energy-efficient appliance you choose, the solar panel you install—are the hands on the churning rope. And every time you choose to mitigate emissions; to contain the poison of our age, you embody the spirit of Shiva.

The gods and demons did not churn the ocean in vain, and neither must we. The *Amrit* of our age lies in clean energy—solar, wind, and hydro. These are the treasures that can sustain life without releasing poison into the air. But they require effort, foresight, and a willingness to invest in the long term. The story of *Samudra Manthan* teaches us that progress is not without peril. It urges us to find balance, to recognise that with every gain comes responsibility. And it reminds us that within each of us lies the power to act—not as passive observers, but as guardians of the world we inhabit.

When you stand beneath the vast sky, with the wind brushing against your face and the sun warming your skin, remember the ocean. Remember the treasures and the poison. Know that your actions are part of a continuum that stretches back to the churning of Mount Mandara.

In every choice lies a whisper of immortality, a chance to leave the world better than you found it.

And as you take your steps, however, small, to reduce carbon emissions, embrace renewable energy, and protect the balance of life, know this: you are not just a part of the story. You are the story. You are the churner, the creator, the preserver, and the one who holds the power to shape the future. The churning continues, and it is ours to guide.

Will you rise to the call?

Will you awaken the Shiva within?

Whisper of the Earth

The Earth speaks in a quiet, persistent whisper, reminding us that every light we switch on and every device we power up contributes to a collective impact.

Embrace renewable energy, cherish efficiency, and let our choices illuminate a path to a sustainable future, for in our hands lies the power to heal or harm. Listen closely, for the Earth's whisper is a call to action, urging us to tread lightly and responsibly.

Beyond the Fence Line

Imagine it is 5 am and you are standing on the balcony of your high-rise apartment gazing at the dawn, and the Earth begins to whisper gently.

These whispers are ancient, laden with secrets of interconnectedness, of cause and effect, of unseen ripples that traverse landscapes and lives.

That is what they say about Scope 3 emissions—a phenomenon not bound by fences or borders but one that threads through our shared existence. A result of pure globalisation.

Imagine, you are picking up a simple affordable cotton t-shirt from a display rack. Uncover its story, not the story of the garment hanging there, but of the journey it undertook to arrive at this very moment.

The cotton was plucked from fields. The threads were spun in a factory that is fired with energy received from a coal-fired power plant. It undergoes ginning, weaving, dyeing, cutting, and shipping processes as part of the various phases along its supply chain. It then crosses continents, is stored in warehouses and is presently in your hands whispering its unspoken tale of emissions.

A story that neither begins nor ends with you. That is the nature of Scope 3 emissions. They are invisible footprints of our modern lives reaching far beyond what the eye can see or what the mind easily comprehends. They are the emissions of suppliers and transporters, customers who use the product, recyclers, and even the emission that rises from landfills. They are everywhere and nowhere, always hiding in plain sight. This unseen emission is a result of modern-day globalisation.

Think of your morning ritual—that daily cup of coffee in your favourite ceramic mug. The beans, cultivated in the mist-laden hills of Colombia. Before they reached the roasters, the beans were washed, sorted, and sundried. These actions seem to be so far removed from your kitchen. The mug itself was born in a fiery kiln in China, crafted from clay that is as ancient as the Earth itself.

Even the act of heating the water for that coffee, carries the whispers of the past; the coal that burned, the dam that held rivers back, the turbine that turned to bring power to your home. These stories, though fragmented and distant, coalesce into the quiet hum of Scope 3 emissions persistent in their reach and scale. Everything contributes towards the energising taste of your morning coffee. The idea is not to discourage you from drinking your morning coffee in your favourite ceramic mug or wearing your cotton t-shirt, but to make you aware of the impact of everything that we use in our daily lives.

Calculating this impact, which is called the Scope 3 impact, is a massive challenge. How do you measure the immeasurable? How do you claim responsibility for what is shared by so many hands? It is a collective puzzle, a narrative from every corner of the Earth.

Yet, within this challenge lies beauty.

Wisdom from the Present

Let's examine the company Patagonia, which specialises in outdoor clothing. A jacket is not just a jacket; it is a pledge. The recycled polyester, the ethical wool, the promises to repair rather than replace—each garment of theirs is a whisper of a different kind—of hope and stewardship. Or let's consider Apple, where sleek design belies the complex partnerships and promises made to suppliers across the globe. The choice to invest in renewable energy projects, to reclaim rare Earth materials, and create a circular economy, is a testament to a vision that sees beyond the bottom line.

Beautiful examples come from individual action. A woman in Mumbai takes a reusable bag to the market, sparing more plastic from ending up into the oceans. A family in Germany opts for an electric car rather than for a car with a growling internal combustion engine. These may not sound like grand gestures, but are tiny revolutions. You buy an electric vehicle and you are a part of that tiny revolution, because you have heard the whisper from the Earth and have responded to it. Simple daily choices are amplified by the power of collective action.

Scope 3 is not just about policies; it is about stories—stories of interconnected lives and choices that bind us all. It asks us to listen to the Earth's whispers to follow the threads of causality that lead from field to factory, from home to horizon. It beckons us to slow down, to see the unseen, to feel the weight of our footsteps on this shared ground. In its quiet complexity, Scope 3 reminds us that no action, however small, is without consequence. The question isn't just whether we can measure these emissions, but whether we are willing to take responsibility for the stories they tell. For in those stories lies the hope of a world that listens, responds, and begins to tread more lightly beyond the fenced line and into the future.

As consumers, you play a vital role. The more you become aware of the environmental impact of your purchasing decisions, the more you can promote eco-friendly practices. The ripple effect starts from you, the consumer. You hold considerable power through your purchasing decisions, lifestyle choices, and by lending your voice and words to this. Your eco-friendly purchasing decisions will have a ripple effect and ensure an increase in the overall demand of more ecofriendly mitigated Scope 3 emission-based product availability. For this to happen, you need to be aware that there is a need for a cultural shift. This shift is driven by increased access to information, greater visibility of climate change impacts, and a broader cultural shift towards sustainability. Adopting more plant-based diets, initially decreasing, and later avoiding, fast fashion trends, carrying refillable water bottles, shopping bags, food containers, and so on, are simple acts that can collectively result in immense environmental benefits.

Imagine for a moment that Scope 3 emissions, the silent whispers, aren't just numbers or graphs but living echoes. These echoes are:

- Echoes of choices
- Echoes of actions

- Echoes of everything that we share as a community; a global community.

These echoes do not belong to globalisation alone; they belong to us all.

Even here, in this light there is a story. The glass in the window came from a sand dune somewhere. The frame, perhaps from a forest long forgotten. The cool swirling air that is wafted down by the fan, is a memory of a river or a hollowed-out coal mine.

This everyday life of each one of us is not a story of a big business. It is your life, it is my life, it is everyone's life. We often hear people saying, 'I am just one person.' It is true, you are just one. But you are a part of a symphony. When you choose to reuse, reduce, repair, or rethink, your voice carries that symphony. A family in Kenya replaces kerosene lamps with solar lights. A college student in Japan swaps single-use plastic for bamboo. A farmer in Brazil plants agroforestry instead of clear-cutting land. A young girl in India prototypes a carry bag made from the shells of a peanut. These acts and choices add up. They resonate in the overall scheme of things. Scope 3 is not a villain born out of globalisation; it is a mirror reflecting how deeply interconnected we have become. It is not here to punish us but to remind us of our connections to each other and to the Earth.

Wisdom from the Ages

Aasha worked tirelessly at her loom in her small village, where life seemed to synchronise with the rhythm of the Earth. Her creations were more than just cloth; they were stories of the world with threads of meaning that transcended borders and cultures. Karen, a traveller, laden with a heavy backpack wandered into her village and stopped outside her house to refill her water bottle. Karen had miles to go, as she was passionate about trekking. Aasha's welcoming nature created a sub-conscious connect and encouraged Karen to take a break.

'It doesn't matter,' Karen thought, 'if I miss today's milestone.' The avid traveller felt it was important to stay and speak to Aasha where she felt connected. They didn't speak one word, but it was just a feeling. Karen decided to stay for a while and set down her pack, sitting on the cemented floor outside the house. Aasha smilingly brought her a glass of water, along with a jug of buttermilk. The women talked about their experiences and how their interest in connecting with nature and hearing its whispers interested them. Karen was more a listener; she spoke less and wanted to hear more.

'What is the secret to harmony?' Karen asked. The question was posed with a sense of restlessness.

Aasha smiled, her hands deftly guiding the shuttle across the warp. She felt the restlessness and urgency in the question. She replied, 'Your question needs a lengthy answer and it can't be provided instantly. You need time and patience to grasp the concept. You need to be here for a while to truly understand this concept.'

Karen nodded to show her interest in learning more, with a smile that displayed her eagerness.

Aasha smiled once again, 'The answer lies in the threads,' she said, tending to each strand in the loom, which would eventually turn into a beautiful shawl. 'Have you heard of *Sumak Kawsay*? It's a Quechua philosophy from the Andes. It means good living. Good living not because of money, but living well by being in harmony with nature and community. Imagine a farmer who tends his fields, not for profit alone, but to nourish the land and his people.

'This is how Earth whispers its gratitude,' Aasha continued, as she added a vibrant green thread to the loom and said, 'this one is for *Friluftsliv*, the Norwegian love for open air. It teaches us to embrace the outdoors, to feel the wind on our faces and understand the joy that comes from being present with the Earth. The red thread here, that is now getting into the warp, is *Swadeshi*, the Indian call for self-reliance and local stewardship. It reminds us to cherish what we create with our own hands, to support our neighbours and to resist the pull of blind consumerism. It's the thread of dignity and purpose. The golden thread goes next, I call it *Sobornost*, the Russian idea of spiritual togetherness and collective responsibility. It is the strength of the community. It reminds us that the burden of care is not yours or mine alone but ours to shoulder together. Imagine a village where no one sleeps hungry because everyone shares what they have. That is Sobornost.'

'Every thread resembling something,' Karen thought. A simple question about harmony, turning into a life lesson. Karen listened with intent, almost rivetted as Aasha continued. The fabric grew richer with every story. 'This deep blue is for *Mottainai*, the Japanese ethos of avoiding waste. It teaches us to respect the resources we use, to find beauty in repair and to cherish what is already ours. Even an old, cracked pot has value if we look with grateful eyes.' Finally, she added a thread of earthy brown. 'This one,' she said, 'is for *Ubuntu*, the African philosophy that says, "I am because we are." It is the heartbeat of humanity, the understanding that our fates are intertwined, like the threads of this cloth.'

'But what ties it all together?' Karen asked.

Aasha's hands paused and she smiled, 'It is *Arete*, the ancient Greek pursuit of excellence in character and purpose. Arete binds these threads with the common goal of living well—not for oneself but for the greater good.' As the loom's rhythm resumed, Aasha turned to Karen, 'These are not lessons that are relics of the past. They are whispers of the present. Whether you live in a crowded city or a remote village, the choices you make—what you buy, what you cherish, how you share—shape the fabric of our world.'

Karen sat in silence, the weight of Aasha's words sinking in. No matter how small the thread, in the grand design of life, it still had something to contribute.

When the shawl was finally complete, it wasn't just a shawl, it contained lessons for life. With each thread speaking a different story, the shawl wasn't a shawl anymore, it was a story of cultures, a living reminder of interdependence, shared responsibility, and a product which resembled the beauty of balance.

It was a pleasant surprise, when Aasha folded it with love and wrapped it in a brown paper and handed it to Karen as a gift. 'Take this with you and spread the story of each thread as you travel across the world,' she said.

Karen, left the village pondering, how a short stop for a water refill became a life-changing lesson with wisdom for her to take away. This frail woman, carried with her immense wisdom from across the world. After all, what was said is more important than who said it. Inside Karen's backpack, was stored that shawl where every thread told a story.

Timeless Tales of Balance

In human history, there exists an unbroken thread, woven through the myths and legends of countless civilisations. This thread binds together ancient stories, their reverence for the Earth, and their pursuit of progress. From the towering cities of Rome, to the sacred waters of the Nile, to the celestial journeys of the Mayas, to the landscapes of Native American lands, humanity has sought not only to understand the world but to shape it. No matter how far we have advanced, one truth has remained steadfast—the balance between nature and progress is delicate and its preservation is the key to our survival. It is this relationship that nurtures the rise of civilisation with the rhythms of the natural world.

Each story, whether it speaks of gods, heroes, or the common folk echoes the same age-old question—how do we progress without losing the essence of the world that sustains us? These stories were not created to entertain—they were crafted to teach, to warn, and to guide. Every civilisation that has risen and fallen throughout history has confronted this question in its own unique way.

The rapid expansion of the Romans left behind smoke-choked skies and scarred landscapes.

The Maya, who once flourished, understood the devastating effects of environmental mismanagement long before the modern world even realised its ecological crisis.

The Silk Road merchants and travellers shared not only goods but wisdom, reminding us that the greatest wealth lies in being in harmony with nature.

Across the globe, stories from Native American cultures, wisdom of the Aboriginal Australians, philosophies of ancient India and China, and ancient myths of Japanese cultures weave a narrative of interdependence.

These are not just fables, but lessons of understanding that the health of the land, sky, and waters is intrinsically linked to the health of humanity itself. In these stories, we see tension between ambition and humility, between human ingenuity, and nature's raw power.

We see the reflection of what we strive to achieve today. How can we advance our technology, our industry, and our society without forsaking the natural world that has always been our foundation?

The civilisations of the past may no longer live among us, but their tales continue to resonate, their wisdom still whispering across time. The aim is to use this section to uncover stories that have stood the test of time; stories that reflect eternal struggles to find balance in a rapidly changing world. Stories that are not reminders, but blueprints of what might still be a tale where each one is a stepping stone on the path towards understanding the dance between the forces of nature and the hands of human progress.

Next, I will share with you more than fifteen short stories from different cultures, and the beauty lies in how each culture, at a high level, stresses the need for balance.

Rome's Smoky Legacy

The Rise and Might of Rome

The mighty Roman Empire, whose influence shaped continents, still echoes in the foundations of the modern world. It was a civilisation that moved forward with unstoppable force. The grandeur of their cities, the marvels of their aqueducts, and the enduring beauty of the Colosseum stand as symbols of human ingenuity and ambition.

Yet, beneath these monumental achievements lay a deeper, more troubling story—one of mindless progress and a tale of unchecked ambition that left scars on the very Earth that sustained it.

Fires of Progress

Rome, in its pursuit of greatness, burned with an intensity that was both literal and metaphorical. The crackling fires that warmed their villas, powered their forges, and fuelled

their industrial enterprises left behind more than just ash. They consumed forests, blackened the skies, and carved a path of destruction throughout the land.

Driven by an insatiable desire to expand, the Romans had little regard for the toll their progress took on nature. The very woods that once whispered in the breeze were felled to feed the fires of their relentless pursuit of power.

A Civilisation that Consumed Instead of Coexisting

This imbalance wasn't found only in the destruction of nature. The empire's vast trade networks, stretching across the Mediterranean and beyond, were fuelled by an endless hunger for resources. Goods flowed in and out of Rome—grain from Egypt, wine from Gaul, olive oil from Spain. Yet, with each shipment, with every extension of their reach, the burden on the Earth grew heavier.

What was meant to be a civilisation of harmony with nature became a civilisation that consumed it. In their zeal for progress, the Romans disrupted the very balance that had allowed them to rise.

Weight of Excess

Rome's legacy is a haunting reminder of the dangers of unchecked ambition. The empire, in all its splendour, failed to recognise the fragility of the world it sought to dominate. Its citizens revelled in the advancements of their time, blind to the quiet devastation that accompanied them.

It is a story of human progress marching forward without regard for the delicate equilibrium of nature—one that, like the empire itself, eventually cracked under the weight of its own excesses.

Lessons for the Modern World

As we look back on Rome's smoky legacy, we are reminded of the fine line between progress and destruction. The ancient Romans may have believed their rise was inevitable, but in hindsight, we can see that their world was built on an imbalance that could not be sustained.

Today, we are at a similar crossroad. In our drive to build, to conquer, to innovate—are we leaving behind the same scars on the Earth that the Romans once did? Rome's story calls us to pause, to reflect on the price of progress, and to recognise that true greatness lies not in how far we can reach, but in how well we can walk in balance with the world around us. If we fail to find harmony, we too may leave behind a legacy of smoke—one that obscures the beauty of the Earth and blinds us to the delicate equilibrium that sustains life.

Modern Takeaway

The story of Rome is a cautionary tale of ambition unchecked by balance. Their relentless pursuit of progress consumed forests, darkened skies, and disrupted ecosystems. The grandeur of their achievements came at the Earth's expense. In the modern world, we stand at a similar threshold. Progress must no longer be measured by expansion alone but by how harmoniously we build with nature. If we fail to learn from Rome's smoky legacy, we risk repeating its mistakes—creating a future where the scars of our ambition overshadow the beauty of the Earth.

True greatness lies in sustainable coexistence, not in conquest.

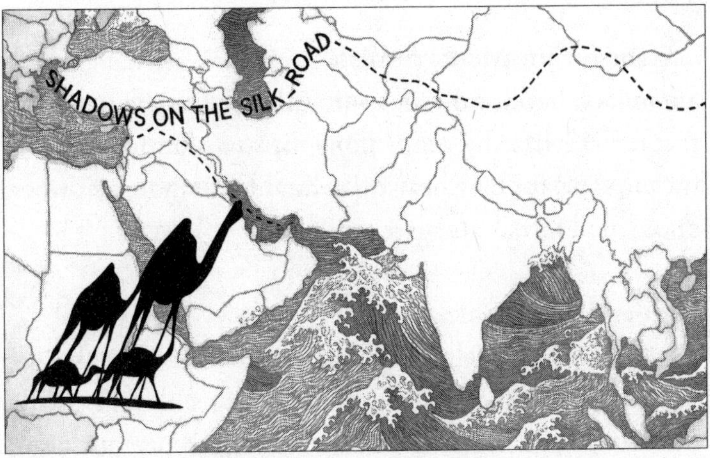

Shadows Left Behind by the Silk Road

The Silk Road, a labyrinthine web of ancient trade routes, was more than just a path for goods; it was a bridge between worlds, cultures, and dreams. It stretched across deserts, mountains, and rivers, connecting the bustling markets of the East to eager traders in the West. Along its dusty trails travelled not only silk, spices, and gold but also ideas, stories, and aspirations—a tapestry of human endeavour woven across thousands of miles.

Yet, beneath the grandeur of its legacy lies a quieter, darker story, one that is often overlooked. This tale is not of wealth or cultural exchange, but of the cost exacted on the Earth by this ancient conduit of globalisation.

Burden of Trade

The caravans, lifelines of the Silk Road, were marvels of endurance, braving harsh terrains and brutal climates. Camels plodded tirelessly under the scorching sun, their hooves kicking up dust that seemed to hang behind like ghosts of their journey. Horses, laden with treasures, carried the burdens of trade and ambition.

These beasts of burden needed sustenance—vast amounts of food and water—draining the fragile ecosystem that they traversed. Camps along the route burned wood for warmth and cooking, leaving behind bare landscapes where once stood forests and greenery.

Hidden Cost of Silk

Beyond the trail itself, the goods that defined the Silk Road bore their own hidden toll. The Chinese silk industry, the heart of this trade, was a marvel of ingenuity but also a voracious consumer of nature's bounty. Vast plantations of mulberry trees stretched across the land, their leaves harvested to feed silkworms. The worms, delicate and demanding, spun threads of unparalleled beauty, but their cultivation generated waste, drained resources, and reshaped landscapes.

The processing of silk—spinning, dyeing, and weaving—was equally resource-intensive, consuming water, fuel, and human labour in staggering amounts.

Scars on the Landscape

As silk, spices, and other treasures made their arduous journey across continents, their passage left an imprint on the Earth. The movement of goods required human effort, animal power, and logistical ingenuity, but also took an invisible toll on the ecosystems it touched. The trails worn into the ground by countless caravans became scars on the landscape, and the water sources that sustained travellers dwindled, their flow diminished by overuse.

The Silk Road was, in many ways, a harbinger of the complexities of modern trade. It exemplified humanity's boundless creativity and ambition but also foreshadowed the environmental costs of our interconnected world. The great irony of this ancient marvel is that while it brought the world closer together, it also underscored how fragile and finite our natural resources truly are.

Lesson for Today

As we reflect on this storied network, we must ask ourselves: how often do we pause to consider the Earth beneath our feet, the rivers that quench our thirst, or the forests that shield us from the sun? The Silk Road, for all its brilliance, is a reminder that progress comes at a price. Its shadows stretch across centuries, urging us to tread more lightly, to honour the balance between human ingenuity and the natural world.

Let the story of the Silk Road serve not just as a testament to human achievement but as a cautionary tale of the unseen costs we leave in our wake. Its whispers ask us to rethink the paths we forge, ensuring that the footprints we leave behind are not scars, but gentle echoes of a harmonious journey.

Modern Takeaway

The Silk Road stands as both a marvel of human ingenuity and a reminder of the environmental costs of interconnectedness. Its legacy teaches us that trade and progress, while transformative, must not come at the expense of the Earth's delicate balance. In our modern, globalised world, where supply chains stretch across continents, the lessons of the Silk Road urge us to consider the true cost of our actions.

Sustainable trade and mindful consumption are not just choices—they are imperatives to ensure that the paths we forge today leave behind not scars, but echoes of harmony with the natural world.

> *Every path we forge leaves a mark—*
> *let it be an echo, not a scar.*

Maya Echoes

Rise of a Civilisation

The Maya civilisation, a beacon of intellect and ingenuity, rose from the dense rainforests of Mesoamerica, leaving behind an indelible legacy. With their awe-inspiring pyramids that seemed to touch the heavens and their profound grasp of celestial movements, the Maya were architects of wonder and explorers of time. Yet, amidst their brilliance lies a narrative of imbalance—a tale of progress that came at a cost to the natural world they so intimately depended upon.

Cost of Monumental Ambition

The limestone pyramids and temples, still standing as guardians of history, were more than architectural feats; they were symbols of the Mayas' ambition and ingenuity. But these grand structures demanded immense labour and resources.

Limestone, abundant yet laborious to process, became the cornerstone of their achievements. Entire forests were felled to fuel the kilns that transformed raw stone into lime, the key ingredient for the plaster that coated their edifices. The smoke from these fires rose to meet the sky, a silent witness of the trees that were turned to ash and the landscapes that were forever altered.

Agriculture: Gift and a Burden

The heart of Maya life pulsed in its fields, where maize was cultivated not just as sustenance, but as a sacred gift from the gods. Vast tracts of forest were cleared to make way for these crops, each swathe of land a promise of life yet a whisper of loss. The relentless expansion of agriculture wore away the soil, leaving it parched and depleted, unable to sustain the very life it once nurtured. The delicate balance of the local climate, already precarious, was disrupted, leading to periods of drought that would challenge the resilience of the civilisation itself.

Trade: A Double-Edged Sword

Trade, the lifeblood of connection and commerce, flourished in the Maya world. From cacao to jade, obsidian to exotic feathers, goods moved along an intricate web of routes, linking the Maya with neighbouring cultures.

These exchanges brought wealth and knowledge, but they also carried a quieter cost. The transportation of goods—whether by canoe gliding through rivers or by foot across jungle paths—required resources that took their toll on the Earth. Each journey etched its mark into the environment, an invisible ledger of the natural world's contributions to human progress.

Slow Unravelling

Yet the Maya, despite their wisdom, could not foresee the long-term consequences of their actions. As their cities grew and their demands on the land intensified, the balance between humanity and nature began to falter.

The forests they cleared, the soil they exhausted, and the ecosystems they altered eventually became their undoing. The collapse of this once-great civilisation was not sudden but a slow unravelling—a poignant reminder of the fragility of harmony.

Echoes of the Maya World

The echoes of the Maya world resonate deeply today. Their story speaks not just of a remarkable civilisation, but also of the choices we make as stewards of our own progress.

In their towering pyramids and abandoned cities, we find lessons etched in stone: that the Earth is generous, but its patience is finite; that progress, unchecked, becomes a burden too heavy for nature to bear. As we look upon the remnants of the Maya civilisation, let us not only admire their achievements but also heed their unspoken warning. The whispers of their forests and rivers, long silenced, call out to us: to tread thoughtfully, to act wisely, and to honour the intricate balance that sustains life. Their story is not just of the past; it is a mirror to our present and a guidepost for our future.

Modern Takeaway

The rise and fall of the Maya civilisation offers a powerful reminder of the delicate balance between progress and nature. While their achievements in architecture, agriculture, and trade were remarkable, their unchecked consumption of resources led to environmental degradation and societal collapse. In our modern world, where technological advancements and consumption are ever-growing, the Maya story urges us to consider the long-term consequences of our actions. Sustainable practices and a deep respect for the Earth's limits are essential to avoid repeating the mistakes of the past. As we build for the future, let us remember that true progress lies in being in harmony with the world around us.

The forests they felled, the soil they drained—
whispers of a past that warn our future:
progress without balance is a path to ruin.

Yudhishthira's Wisdom: Lesson for Our Times

In the sacred forests of ancient India, where the whispers carried wisdom of ages, a story unfolded—a tale that speaks to the heart of human responsibility and the delicate balance of life. This is the story of Yudhishthira, the eldest of the Pandava brothers, whose wisdom illuminated a path through the darkness of ignorance.

Exiled from their kingdom, the Pandavas wandered

through a desolate wilderness. One fateful day, parched under the blazing sun, they came upon a serene lake. But this was no ordinary lake—it was guarded by a *Yaksha*, a mystical being who imposed a test. Anyone who sought its waters had to first answer a series of riddles. Those who failed lost their lives. One by one, Yudhishthira's brothers approached the lake, dismissing the Yaksha's questions in their haste. One by one, they succumbed to the Yaksha's curse, their bodies lying still by the water's edge. Finally, Yudhishthira arrived, weary yet resolute. Faced with the same challenge, he paused—his heart heavy with grief, but his mind unwavering.

The Question that Echoes Through Time

- The Yaksha's questions were not mere puzzles but reflections on life, death, and the unseen threads that bind existence.
- 'What is the greatest wonder in the world?' the Yaksha asked, his voice like the rustling of ancient leaves.

Yudhishthira answered:

- 'It is that people see others dying every day, yet live as if they are immortal.'
- The Yaksha, moved by the depth of his response, restored the fallen brothers to life. The Pandavas continued their journey, humbled and enlightened.

Whisper of the Earth: Our Modern Reflection

Yudhishthira's answer is a mirror to our own lives. We see the signs of nature's fragility—rising seas, dwindling forests, and shifting climates—yet we live as if these changes are distant, as if the Earth's resources are infinite. Like the brothers who ignored the Yaksha's warning, we charge forward, heedless of the consequences. The lake, serene yet perilous, symbolises our planet—a source of life that demands wisdom and restraint. The Yaksha's questions are the unspoken inquiries of nature itself, urging us to pause, reflect, and act responsibly. Deforestation, fossil fuel consumption, and the unchecked exploitation of resources are all whispers of imbalance. Just as Yudhishthira saw beyond the illusion of permanence, we too must recognise that our world is not limitless.

Call to Wisdom and Balance

Yudhishthira's wisdom lay not just in answering the questions but in understanding the interconnectedness of life. True progress is not measured by the monuments we build or the wealth we amass but by the care we take to sustain the balance of life. Let his story guide us. Let his wisdom remind us that the legacy we leave must be one of stewardship, not depletion. Only through foresight and humility can we ensure that future generations inherit a world still rich with life.

Modern Takeaway

Despite our technological progress, we continue to live under the illusion of immortality—believing that the Earth's resources will last forever, that the damage we do will not affect us in our lifetime. But the Earth's balance is delicate, and our actions are leaving irreversible scars. Just as Yudhishthira recognised the greatest wonder—that humans ignore the inevitability of death—so too must we acknowledge that nature is not invulnerable. The Yaksha's challenge is our challenge: to pause, reflect, and act before it is too late. If we do not heed this wisdom, we may wake up one day, like the Pandavas at the lake, only to realise that we have lost what truly sustains us.

> *You see the forests falling, the rivers drying, yet you live as if the Earth is immortal.*

Tale of Yu the Great

In ancient China, the land was constantly ravaged by floods. Rivers overflowed, destroying fields and homes. The people were helpless, unable to stop the water's destructive force. But then came Yu, a man determined to bring peace to the rivers. Unlike others who tried to fight nature, Yu sought to understand it. He didn't build barriers, but dug channels for the rivers to flow through. He created paths that allowed the water to nourish the land rather than destroy it. Yu spent years working tirelessly, designing irrigation systems that mimicked the natural flow of water. He didn't rush, but carefully observed how the rivers shaped the land and how the soil supported life. His approach brought people together. They dug canals, built terraces, and planted trees.

Over time, the floods receded, and the land began to heal. Fields flourished, villages were rebuilt, and the rivers flowed in harmony with the land. Yu's story teaches us that progress doesn't come from trying to control nature, but from working with it. Today, we face our own challenges with the environment. Like the rivers that once flooded China, our actions—if unchecked—can harm the Earth. But if we understand the interconnectedness of nature and act thoughtfully, we can create solutions that work for both the planet and humanity. Yu's wisdom reminds us that true progress is not about quick fixes but long-term harmony. His legacy challenges us to think beyond short-term gain and to act with respect for the Earth's natural balance.

Modern Takeaway

In today's world, Yu's approach offers a powerful lesson: progress doesn't come from forcing nature to bend to our will but from understanding and working with it. As we face environmental challenges like climate change, resource depletion, and pollution, we must remember that sustainable solutions are those that respect the delicate balance of the natural world. Just as Yu harmonised the rivers with the land, we must find ways to create systems that support both human needs and environmental health. This means considering the long-term impact of our actions, using resources wisely, and fostering cooperation rather than dominance over nature. Yu's legacy calls us to build a future where both the planet and its people thrive in harmony.

Not by damming, but by listening—
the flood yields to wisdom.

Story of the Earth Divers

In the primordial waters, before Earth as we know it, came into being, there was nothing but an endless sea. The sky stretched above, a dome of infinite potential, but beneath it, there was no land, no home for life to thrive. The world was adrift, waiting for a moment of creation. The Great Spirit, observing this emptiness, sought to bring balance and life to the universe. But the Spirit could not do it alone. The task required more than divine will—it needed the courage, determination, and unity of all beings.

And so, a call went out to the animals dwelling in the watery depths and the boundless skies: 'Who among you will dive into the depths and bring forth soil to create the

land?' The animals looked at one another, hesitant but hopeful. It was an immense task, for the waters were deep and dark, and the soil was buried far below. Yet, one by one, they volunteered.

The graceful Loon was the first to dive, its sleek body cutting through the water. Down it swam, its wings moving with purpose, but the depth was too great, and the Loon returned empty-handed, gasping for air. Next came the powerful Otter, whose strength and agility were unmatched. It dove with vigour, its heart set on the mission. But the darkness below was overwhelming, and the Otter, too, surfaced with nothing but exhaustion. One by one, the animals tried—the determined Beaver, the wise Heron—but all met with the same fate. The soil seemed unreachable, and doubt began to creep in. Perhaps the task was too great, the dream too far.

Then, a small voice broke the silence. It was the humble Muskrat, often overlooked for its size and strength. 'I will try,' it said, its voice steady with quiet determination. The others watched in disbelief, unsure if such a small creature could succeed where the mighty had failed. The Muskrat dove into the waters, its tiny form disappearing into the abyss. The darkness closed in, and the pressure grew, but the Muskrat pushed on, its heart beating with a singular purpose. Time seemed to stretch endlessly. Just when the others feared it had been lost, the Muskrat emerged, its fur soaked and its body trembling. In its tiny paw, it held a clump of soil—a gift from the depths, the seed of creation. The animals cheered, their spirits lifted by the Muskrat's courage.

The Great Spirit took the soil and placed it gently on the back of the Great Turtle, who had volunteered to bear the weight of the new land. Slowly, the Earth began to grow, spreading across the turtle's broad shell, a home for all beings. The story of the Earth Divers is more than a tale of creation—it is a testament to the power of collaboration and collective responsibility.

Each animal played a role, their efforts weaving together the fabric of life itself. The smallest among them, through sheer will and determination, proved that no contribution is too small when the goal is shared.

This ancient legend whispers to us across time, urging us to remember that the world we inhabit is not ours alone. Every action we take, every choice we make, ripples through the interconnected web of life. Like the animals who worked together to create the Earth, we too must unite in the face of the challenges that threaten our shared home.

Whether it's in how we use resources, the ways in which we produce and consume, or the industries we build, the lesson is clear: sustainability is not an individual endeavour. It requires the cooperation of all—nations, businesses, communities, and individuals alike. Just as the animals dove into the unknown for the sake of creation, we are called upon to act with courage and unity, forging a path that honours the balance of life. May the story of the Earth Divers remind us of our shared responsibility to nurture the land we walk on, the waters that sustain us, and the sky that watches over us. Let us remember that even the smallest effort, when joined with others, can shape the future of our world.

Modern Takeaway

In today's world, the story of the Earth Divers serves as a reminder that no contribution is too small when it comes to sustainability. Just as the animals united to create the Earth, we must collaborate across all sectors—governments, businesses, and individuals—to address the challenges facing our planet. Whether it's reducing waste, conserving resources, or supporting sustainable practices, every action counts. Just as the Muskrat's small but determined effort changed the course of creation, our collective efforts, no matter how small they seem, can have a profound impact on the future of our world.

> *The effort of even the humblest one of us can create a world of difference.*

Anansi and the Weight of Wisdom

Under the dappled shade of an ancient Baobab tree, Anansi the Spider sat weaving his web, his eight legs moving with a rhythm that mirrored the harmony of the world around him. The sun, casting its golden light, seemed to whisper to him, reminding him of the countless lessons hidden

in the land, rivers, and skies. Yet, Anansi's thoughts were elsewhere. He was consumed by a single desire—to gather all the wisdom in the world and claim it for himself.

Anansi believed that with such wisdom, he could master everything. He would control the seasons, dictate the growth of the crops, and even command the rains. He envisioned himself as the one who held the answers, the solutions, and the power to shape the world as he pleased. And so, he began his journey. He travelled across hills, valleys, and rivers, collecting wisdom wherever he found it—how the rain nourished the Earth, how the wind carried seeds to new homes, how the ants worked tirelessly in unity. Each fragment of knowledge was stored in a great gourd, sealed tightly to ensure that no one else could access it.

But as the gourd grew heavier with each new piece of wisdom, Anansi noticed something peculiar. The rivers he had admired began to dry up, their waters unable to sustain the life they once nurtured. The fields he passed grew barren, the soil cracked and unyielding. The harmony he had taken for granted was faltering, as if the balance of the world depended on the very wisdom he was hoarding.

Determined to protect his treasure, Anansi decided to hide the gourd in the tallest tree. He tied the gourd to his back and began to climb. But the weight of it dragged him down, and time after time, he tumbled to the ground, exhausted and frustrated. As he sat beneath the tree, pondering his failure, his young son Ntikuma approached. 'Father,' Ntikuma said softly, 'why don't you tie the gourd to your belly instead of your back? It will balance you better.' Anansi froze. The simplicity of the suggestion stunned him. Despite all the wisdom he had gathered, he had overlooked this basic idea.

In that moment, he realised a truth: wisdom, like the resources of the Earth, is not meant to be confined or hoarded. It grows in value only when it is shared. With a laugh, Anansi untied the gourd and opened it, letting the wisdom flow out into the world. It scattered like seeds on the wind, spreading to every corner, enriching the lives of all who encountered it.

Anansi's story offers a timeless lesson about the dangers of greed and the power of collective responsibility. His initial actions mirror humanity's unchecked exploitation of natural resources—hoarding, consuming, and depleting without consideration of the broader consequences.

The drying rivers and barren fields in the story are reminders of the environmental imbalance that follows when resources are misused for personal gain. But Anansi's epiphany is a call to action. It speaks of the need for transparency, collaboration, and a mindset that prioritises the greater good. Just as Anansi learned that wisdom grows when shared, we must recognise that sustainable practices, innovative solutions, and environmental stewardship must be accessible to all.

This story challenges us to think beyond individual gain and embrace a vision of shared responsibility. It reminds us that the interconnectedness of our actions shapes the health of the world. Like Anansi's scattered wisdom, our efforts must ripple outward, creating a legacy of balance, harmony, and a better future for generations to come. By releasing the weight of selfish ambition and choosing the path of sharing and collaboration, we can rewrite our story—one where wisdom and resources are not burdens, but gifts for a flourishing Earth.

Modern Takeaway

The story of Anansi reminds us that wisdom and resources should never be hoarded. In today's world, the unchecked exploitation of natural resources and the pursuit of personal gain without regard for the greater good lead to environmental imbalance. The modern takeaway is clear: sustainability thrives when knowledge, resources, and efforts are shared. Collective responsibility, transparency, and collaboration are key to creating a future where both people and the planet flourish. By prioritising the common good over individual gain, we can build a balanced, harmonious world for future generations.

Wisdom, like nature, thrives only when shared.

Bamboo Cutter: Legends of Grace and Gratitude

In ancient Japan, among rolling hills and whispering forests, lived an old bamboo cutter named Taketori. Each day, he walked through the bamboo groves, his hands caressing the smooth stalks as he chose which to harvest. Taketori took only what he needed, never more, for he understood the rhythm of the Earth and the silent promise it made to those who tread lightly upon it. One golden morning, as the sunlight filtered through the bamboo canopy, Taketori's blade struck something unexpected. Inside a glowing stalk, he found a tiny, radiant figure—a princess, no larger than his hand. Her presence filled the grove with an otherworldly light. Overwhelmed by awe and joy, Taketori carefully carried her home.

The princess, whom he named Kaguya, brought untold blessings to Taketori and his wife. With her guidance, their modest life blossomed into abundance. The once-dry fields yielded bountiful crops, the streams ran clearer, and the air seemed imbued with a rare vitality. Word of their prosperity spread, and soon, the villagers, too, began to flourish under her quiet influence. But as the seasons turned, Kaguya grew wistful. One evening, beneath a sky painted with the hues of dusk, she revealed her secret. She was a celestial being, sent to remind humanity of the fragile beauty of the Earth. Her time among mortals was drawing to an end.

Her departure was bittersweet. The villagers, once consumed by their newfound wealth, now understood the weight of her parting words: 'All that the Earth gives is a gift, not a possession. Care for it, and it will care for you. Forget this, and you risk losing all.' With tears in their eyes, they watched as she ascended to the heavens, her light fading into the stars. The tale of Princess Kaguya carries lessons about the relationship between humanity and the natural world. The bamboo cutter's life represents a harmonious coexistence with nature, where resources are used thoughtfully and sparingly. Kaguya's blessings, though bountiful, come with a caveat: prosperity that disregards the limits of nature is fleeting.

In today's world, where rapid industrialisation and consumerism have stretched the Earth's resources to their breaking point, Kaguya's message resonates. Industries and individuals alike must recognise the finite nature of these resources and embrace sustainable practices. Overharvesting, unchecked consumption, and the pursuit of short-term gains mirror the initial joy of the villagers—blinding us to the long-term consequences.

Just as Kaguya's departure left the villagers with a renewed sense of stewardship, our choices today must be guided by the understanding that true prosperity lies not in what we take, but in how we preserve. By respecting the delicate balance of ecosystems and fostering collaboration, we can ensure that the blessings of the Earth endure for generations. Kaguya's story, like the gentle glow of the moon, is a reminder: the Earth's gifts are luminous and fleeting, meant to be cherished and shared. In protecting them, we honour both the beauty of nature and the grace of its silent benefactors.

Modern Takeaway

The story of Princess Kaguya teaches us that the Earth's resources are not ours to possess, but to cherish and protect. In a world where overconsumption and environmental degradation threaten our future, Kaguya's message is clear: true prosperity comes from living in harmony with nature.

Sustainability is about understanding limits, embracing stewardship, and fostering practices that honour the delicate balance of ecosystems. Just as the villagers learned too late, we must act now to preserve the Earth's gifts for future generations. Only by respecting nature can we ensure its blessings endure.

> *What we cherish with care will endure,*
> *what we claim as a possession will fade.*

Santa Claus and the Spirit of a Greener Christmas

The tale of Santa Claus is one that stirs the heart. For generations, it has brought joy, warmth, and wonder to children and adults alike. There's something magical about the thought of a jolly figure riding through the frosty skies, guided by reindeer, carrying a sack brimming with gifts for the world's children. Santa's story transcends cultures

and beliefs, weaving together a universal tapestry of hope, compassion, and generosity.

As children, we wrote letters to Santa, whispering our wishes into the frosty air, believing that somewhere in the snowy expanse of the North Pole, Santa would hear. The gifts under the tree were more than toys—they were tokens of love and a reminder that, in a vast and sometimes daunting world, there was still magic to be found.

But as we grew older, reality tempered this magic. We learned that Santa was a story—a beautiful, enduring tale meant to nurture the goodness within us. Yet, even in understanding this, we never stopped wishing for a world where such unselfish generosity was real. Perhaps, in a way, Santa *is* real—not in form, but in the spirit he represents.

Santa's Workshop: New Chapter in the Legend

In the heart of the Arctic, where snowflakes swirl and the auroras dance across the endless night, Santa's legendary workshop has always been portrayed as a hive of cheerful activity. But in recent years, the elves have faced a sobering truth: the ice that cradles their magical home is melting. The very foundation of their world—and ours—is shifting beneath their feet.

Moved by the urgency of the moment, Santa and his loyal elves embarked on a profound transformation. They realised that their workshop could no longer exist as it had in tales of old. It needed to become a beacon of sustainability, a place where magic and mindfulness entwined.

The elves, with their boundless creativity, took to the challenge with gusto. They reimagined their toy-making processes, replacing harmful materials with biodegradable and recycled alternatives. The hum of the workshop was now powered by renewable energy, with windmills turning gracefully against the icy horizon and solar panels gleaming in the pale winter sun.

Even Rudolph and his fellow reindeer found their roles evolving. Their sleigh rides now carried messages as important as the gifts themselves—stories of conservation, kindness, and the power of small actions to create big change.

A Gift Beyond the Wrapping Paper

Each gift that Santa delivers now carries an invisible but strong message: *Cherish the Earth*. The wooden train sets, crafted from sustainably harvested timber, whisper of forests that continue to thrive. The dolls, stitched with care, reflect a promise to preserve the resources that make their creation possible. The puzzle pieces, carved from recycled materials, fit together like the delicate balance of ecosystems that sustain life. Santa's reimagined mission is not about giving less—it's about giving *better*. He reminds us that the truest gifts are those that ensure a brighter future: clean air, clear waters, and a planet that thrives for generations to come.

Santa's Message to the World

Santa's journey has inspired countless communities. Families are finding joy in simpler, more sustainable holidays, exchanging handmade gifts or experiences instead of excess. Cities are embracing greener practices, drawing inspiration from Santa's example to reduce emissions, conserve energy, and protect biodiversity. Santa's sleigh, once a symbol of pure holiday magic, now carries a deeper significance. It soars across the globe, a testament to the power of collective action. The twinkling lights of Christmas remind us not just of celebration, but of our responsibility to ensure that the Earth's own lights—its forests, oceans, and skies—continue to shine.

Eternal Spirit of Santa

Even as the ice cracks and the world faces mounting challenges, the spirit of Santa endures, evolving with the times. He teaches us that generosity is not confined to material wealth; it extends to the care we show our planet and the legacy we leave for those who follow. Perhaps the most magical gift Santa has ever given is this: a reminder that each of us, in our own way, can be a part of this transformation.

We can write a new letter, not to Santa, but to ourselves—a promise to tread lightly, to give thoughtfully, and to ensure that the wonder of this world remains for those who come after us. Santa Claus is no longer just a mythical figure; he is a symbol of what we can achieve together. In the echo of his hearty laugh and the soft jingle of his sleigh bells, we hear the call to a greener, kinder, and more sustainable world.

Modern Takeaway

Santa's reimagined mission teaches us that true generosity goes beyond material gifts—it's about giving back to the Earth. In a world facing environmental challenges, we must embrace sustainability in all aspects of life.

Just as Santa's workshop transformed to reflect eco-conscious choices, we too can make mindful decisions to preserve our planet for future generations. The greatest gifts we can give are those that ensure a healthier, more sustainable world—a world where magic, kindness, and care for the Earth thrive together.

The true gift of Christmas is a lighter footprint on the world we share.

Dewi Sri: Rice Guardian and Keeper of Balance

Gift from the Heavens

In the heart of the Indonesian archipelago, where rice paddies stretch like emerald carpets, the legend of Dewi Sri endures. She is not just a goddess of rice and fertility—she is a symbol of life itself. Sent from the heavens, Dewi Sri carried the wisdom of nature's harmony. She whispered her teachings to the farmers beneath the tropical sun:

Respect the soil, honour the water, and cherish the seeds.

Blessing of Abundance

Under her guidance, barren fields flourished. Golden grains swayed in the breeze, and villages thrived. The people cultivated their land with care, knowing that prosperity came from balance. Her lessons extended beyond farming. She warned of the perils of greed and exploitation:

'Treat the Earth as you would a friend. If you harm her, you harm yourself.'

For a time, the people listened. They celebrated their harvests with gratitude, honouring Dewi Sri through music, prayers, and festivals.

Withering Fields

But as the years passed, greed took root. Forests were cleared without thought, waters were polluted, and the land was overworked. In pursuit of more, they forgot the goddess's teachings. The fields withered, the soil grew barren, and famine spread. Dewi Sri withdrew her blessings, leaving only silence. The people cried out, but the Earth had spoken—*abundance cannot thrive in imbalance.*

Return to Wisdom

In their sufferings, they remembered. Slowly, they restored the land, planting trees, conserving water, and returning to the sustainable practices Dewi Sri had taught. Their efforts were rewarded. The rains returned, the fields blossomed, and the goddess's blessings graced the land once more. Today, Indonesian farmers continue to honour these lessons, preserving biodiversity and nurturing the soil.

Lesson for the World

Dewi Sri's story carries a universal truth—*our choices shape the balance of life.* Whether conserving water, reducing waste, or protecting forests, every action ripples through the web of existence. Industries today strive to produce more with less, but her story is a reminder that *exploitation leads only to emptiness.* Sustainability is not a sacrifice—it is the path to true abundance.

Dewi Sri lives on, not just in Indonesia's rice fields but in the hearts of all who seek harmony with nature. Her voice lingers in the rustling grains, calling us to *give as much as we take and ensure that the Earth's gifts endure for generations.* As the sun sets over the paddy fields, casting them in a golden glow, her whisper remains.

Prosperity is not in how much we take, but in how well we give back.

Kami: The Spirits of Nature and the Soul of Japan

In Japanese mythology, the world is alive with spirits—sacred beings known as *Kami*. These are not distant deities but intimate guardians, inhabiting rivers, trees, mountains, and even the winds that sweep through the valleys. The Kami embody the soul of nature, safeguarding its balance and sanctity. Their story is one of creation, connection, and caution—a timeless reminder that humanity and nature are deeply intertwined.

Creation of Japan: Tale of Love and Loss

Long ago, the celestial gods sought to bring order to the chaos of the nascent Earth. They entrusted this mission to Izanagi and Izanami; divine beings filled with grace and purpose. Standing on the floating bridge of heaven, Izanagi dipped his jewelled spear into the formless waters below. As he lifted it, droplets fell, solidifying into the first island—Onogoro. From this foundation, the couple created the islands of Japan, each infused with the essence of Kami.

Yet, creation came at a cost. Izanami perished while

giving birth to the fire deity, her death fracturing the harmony they had cultivated. Overcome with grief, Izanagi descended into Yomi, the underworld, to bring her back. But the darkness had consumed her. Unable to bear her transformation, he fled, leaving behind sorrow that rippled across the world—manifesting as storms, earthquakes, and floods.

This tale serves as a caution: disrupting the natural order invites chaos and suffering. It is a lesson woven into the Japanese reverence for Kami and their enduring commitment to harmony with nature.

Sacred Spaces: Where Kami Reside

Across Japan, Shinto shrines stand in the heart of ancient forests, beside cascading waterfalls, and atop mist-covered mountains. These are not merely places of worship but sanctuaries of the Earth itself. Each shrine honours a Kami, inviting visitors to pause, reflect, and reconnect with the natural world. To stand before a shrine is to feel time slow—to hear the whispers of the wind through sacred groves.

Prayers are offered, not just in words but in the way in which people respect the land, bowing in gratitude to the spirits that dwell within every stone, leaf, and stream.

Kami in Everyday Life

The reverence for Kami extends beyond shrines into daily life. Farmers work in tune with the seasons, mindful of the land's needs. In the *Satoyama* landscapes—where human life and nature exist in harmony—villagers carefully manage forests, sustain biodiversity, and nurture rice paddies alive with dragonflies and frogs.

Even in Japan's cities, the spirit of Kami thrives. Rooftop gardens bloom, green belts soften industrial zones, and eco-friendly architecture reflects an ancient understanding: to care for nature is to secure the future.

Kami's Warning: A Call to Action

The tragedy of Izanagi and Izanami warns that imbalance—whether through grief, greed, or neglect—has consequences. Today, as the Earth faces deforestation, pollution, and climate change, the warnings of the Kami grow louder.

Rivers, once sacred lifelines, are choked with waste. Skies, once pure, are clouded with emissions. But just as the past holds warnings, it also holds wisdom. Japan's conservation efforts—rooted in Satoyama traditions and modern sustainability initiatives—show that ancient knowledge can guide contemporary solutions.

Spirit of Kami: Global Lesson

The message of the Kami is universal. It speaks to the interconnectedness of all life and the responsibility we bear to nurture the world we inhabit. Every action—planting a tree, reducing waste, or choosing sustainable energy—reverberates through the web of existence.

To honour the *Kami* is to honour the Earth. It is to recognise that our well-being depends on the health of the planet. As we navigate an uncertain future, may we carry the spirit of Kami in our hearts, treating every forest, river, and mountain as sacred, and every act of preservation as a prayer. For the Kami are watching, not as distant overseers, but as quiet whispers in the wind, urging us to tread lightly, to live deeply, and to listen.

The Kami are not lost to time—they live in every leaf that rustles, in every river that runs, waiting for us to listen.

Lord Ganesha: Keeper of Harmony, Guardian of Nature

Lord Ganesha stands as an enduring symbol of wisdom, balance, and resilience. His elephant-headed form is not just a matter of divine imagery but a profound metaphor for our relationship with the natural world. Elephants have long been revered in Indian culture, not just for their strength and grandeur but for the role they play in shaping and sustaining ecosystems. As they move through forests, they create pathways for other creatures, disperse seeds that give rise to new life, and maintain the delicate balance of the habitat.

Ganesha, with his large ears and thoughtful gaze, embodies this wisdom—listening deeply, observing keenly, and acting with both strength and sensitivity. His form is a reminder that nature speaks in subtle yet powerful ways. The rustling of leaves, the steady rhythm of flowing rivers, and the silent resilience of mountains—each carries a message,

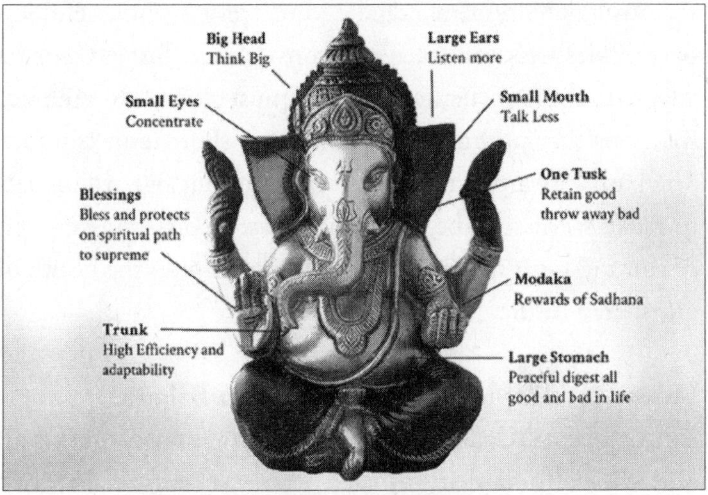

a whisper of the Earth that urges us to pay attention, to act with care. To protect these whispers is to protect life itself.

Broken Tusk: Lesson in Sacrifice and Sustainability

One of the most evocative stories of Ganesha is his role as the scribe of the Mahabharata. When Sage Vyasa set out to dictate the great epic, he sought a writer who could match the speed and intensity of his thoughts. Ganesha agreed, on the condition that Vyasa would not pause in his recitation. As the words poured forth in an unbroken flow, Ganesha's pen broke. But instead of stopping, he broke off one of his tusks and continued writing, undeterred by loss or discomfort.

This act of sacrifice speaks of a fundamental truth—commitment to a greater purpose often requires letting go of something we hold dear. In today's world, where excess consumption and wastefulness have become the norm, Ganesha's broken tusk stands as a powerful reminder: true progress is not about relentless acquisition but about knowing what to give up for the sake of balance.

Sustainability demands the same spirit—resourcefulness over recklessness, resilience over convenience. Just as Ganesha adapted without hesitation, we must learn to embrace solutions that prioritise the planet over short-term comfort. The path forward is not always smooth, but our willingness to make sacrifices—be it reducing waste, conserving energy, or shifting to sustainable practices—determines the health of our shared future.

Mouse and Elephant: A Partnership in Balance

At first glance, Ganesha's vehicle—a tiny mouse—seems an odd choice for a deity of such immense stature. Yet, this pairing carries a lesson that is timeless. The elephant and the mouse exist at opposite ends of the spectrum, one towering, the other small; one slow and measured, the other swift and restless. And yet, they move together in harmony, each compensating for the other's limitations.

The mouse represents agility, persistence, and the ability to navigate through the smallest of spaces—qualities that balance the elephant's strength and steadiness. This unlikely partnership is a metaphor for ecological equilibrium. Every being, no matter how big or small, plays a role in maintaining the balance of life.

Today, as human activity disrupts natural order—deforestation stripping away habitats, unchecked emissions altering the climate, pollution choking rivers and air—this lesson is more urgent than ever. If one part of the ecosystem suffers, the entire web of life is affected. The bond between Ganesha and his mouse is a call to rethink our relationship with nature, not as its masters, but as cohabitants in an intricate and interdependent world.

Ganesh Chaturthi: A Ritual of Reverence

Every year, the festival of Ganesh Chaturthi transforms India's streets and homes into vibrant centres of devotion. Clay idols of Ganesha, lovingly sculpted and adorned, take centre stage in celebrations filled with music, prayers, and joy. Yet, the most poignant moment comes at the festival's end, when these idols are immersed in water, dissolving back into the rivers, lakes, and oceans. This cycle of creation and dissolution mirrors the natural order—everything that arises must eventually return to the elements.

However, in recent years, the spirit of this ritual

has been challenged by the environmental impact of non-biodegradable idols and chemical-laden paints. As awareness grows, many communities are returning to traditional, eco-friendly practices—using clay, natural dyes, and sustainable materials to ensure that devotion does not come at the cost of nature.

This shift is more than a practical change; it is a reaffirmation of the very essence of Ganesha's teachings—respect for the Earth, mindfulness in action, and harmony between tradition and responsibility. It reminds us that rituals, no matter how sacred, must evolve in alignment with the well-being of the world they honour.

Ganesha's Teachings in the Modern World

In an era of climate change, depleting natural resources, and environmental instability, Ganesha's stories hold a renewed significance. His wisdom speaks to the urgency of our times.

Just as he removes obstacles from the paths of his devotees, we must remove the barriers that hinder sustainable progress—whether it is resistance to clean energy, unsustainable industrial practices, or the mindset of limitless consumption.

His large ears remind us to listen—not just to words, but to the silent warnings of a changing climate, vanishing forests, and endangered species.

His broken tusk is a call to sacrifice extravagance for necessity, to let go of excess in favour of sufficiency.

Ancient wisdom does not exist in isolation; it is a guidepost for modern solutions. From community-led afforestation projects to innovations in renewable energy, the principles Ganesha embodies can shape the way we approach sustainability. The question is not whether mythology and science can coexist, but how they can come together to create a future where both tradition and progress thrive.

Soulful Connection to the Earth

To gaze upon Ganesha is to feel a deep connection to the world around us. His stories are not mere fables; they are blueprints for a life lived in harmony with nature.

Every choice we make—turning off an unnecessary light, reducing waste, planting a tree—carries significance beyond our immediate surroundings. Just as a river shapes the land it flows through, our actions ripple outward, affecting communities, ecosystems, and generations to come.

Ganesha's broken tusk reminds us that every meaningful endeavour requires sacrifice. His partnership with the mouse teaches that no challenge is too great when approached with creativity and determination. His elephantine presence calls us to listen, not only with our ears but with our hearts—to the Earth, to its creatures, and to the wisdom embedded in nature itself.

As the world stands at a crossroads, facing rising emissions and dwindling resources, Ganesha's message is clear: sustainability is not an act of convenience but a commitment to balance. The choice before us is not whether to act, but how swiftly and wisely we choose to do so.

Harmony is not about dominance or submission; it is about understanding that every being, every element, and every moment is part of a delicate balance worth preserving.

The Fisherman and the Djinn: Soulful Reflection

Cast into Fate

From the mystical sands of the Middle East comes a tale whispered across centuries—a story not just of adventure and cunning but of harmony with the forces that shape our world. Nestled within the timeless *Arabian Nights*, the story of the Fisherman and the Djinn unravels profound truths about our relationship with nature—a bond as delicate as the threads of a fisherman's net.

A humble fisherman rises before dawn, his calloused hands bearing witness to a life spent on the edge of survival. Each cast of his net into the vast, unpredictable sea is a plea for sustenance, a silent prayer to the waters that have long been both his benefactor and his bane. But on this day, the sea seems reluctant. The first cast brings only weeds. The second, broken shells. The third, a pitiful cluster of bones.

His heart sinks; his family will go hungry.

On the fourth attempt, his net emerges heavy, promising—but instead of fish, it holds a sealed brass jar, encrusted with the barnacles of time. Intrigued, the fisherman opens it, unleashing a whirlwind that darkens the sky. A mighty Djinn, towering and tempestuous, bursts forth, his voice booming like a storm. The fisherman stumbles back, terror coursing through him as the Djinn proclaims his intent to kill the one who freed him, his wrath honed over centuries of confinement.

Wrath of the Djinn

The Djinn's anger is not unfounded. He recounts his story—once a force of nature, governing the seas and winds, he had been a steward of balance. But his rebellion against the divine order led to his imprisonment by King Solomon, who cast him into the jar and hurled it into the depths of the sea. Over millennia, the Djinn's rage festered, turning him into a force of destruction rather than harmony.

The fisherman, though trembling, is clever. He feigns disbelief, taunting the Djinn: how could such a mighty being fit into so small a vessel?

The Djinn, prideful and eager to prove his might, transforms into smoke and funnels back into the jar. With a swift motion, the fisherman seals it, his hands shaking but resolute.

A Pact with the Storm

The ensuing dialogue between man and Djinn is laden with tension but also with hope. The fisherman bargains for his life, offering to release the Djinn if he promises to abandon his vengeance. The Djinn, touched by the fisherman's courage and wisdom, agrees, and their pact is sealed with a solemn understanding.

As the fisherman releases the Djinn, they part not as adversaries, but as partners in a fragile truce. The sea, once again calm, mirrors the possibility of balance restored. The fisherman returns home, his net empty, but his heart full—he carries a lesson far greater than the day's catch.

A Parable for the Present

This ancient tale, though simple on the surface, brims with lessons for the modern world. The Djinn, once a symbol of balance and power, reflects the natural forces that sustain our planet—rivers that carve life into parched lands, winds that carry seeds across continents, and oceans that cradle the rhythm of life. But when these forces are exploited or disregarded, they can turn destructive, much like the enraged Djinn.

The fisherman's plight mirrors humanity's relationship with nature. His initial harmony with the sea, taking only what he needs, speaks of an equilibrium long lost in our age of excess. But the jar, dredged from the ocean floor, is a harbinger of disruption—a symbol of humanity's ceaseless extraction of resources, polluting and disturbing ecosystems that have thrived for millennia.

The Djinn's fury can be likened to the emissions we release into the atmosphere, the scars we leave on forests, rivers, and oceans. It is a warning that unchecked actions can awaken forces beyond our control, forces that will not discriminate between the innocent and the guilty.

Restoring the Balance

Yet the tale does not end in despair. The fisherman's courage and ingenuity pave the way for reconciliation. This reflects our potential to mend what has been broken, to heal the wounds inflicted on our planet. Like the fisherman striking a pact with the Djinn, humanity too can find ways to coexist with nature, channelling its power for good rather than harm. It calls for wisdom, respect, and a recognition of our shared fate with all life forms.

The story of the fisherman and the Djinn whispers a timeless truth: that the forces of nature, though vast and untameable, are not our enemies. They are allies—if only we honour the delicate dance of coexistence. In a world grappling with the consequences of climate change, deforestation, and rising emissions, this story offers a beacon of hope: that even when the storm rages, reconciliation and harmony are within our grasp. If only we listen. If only we act.

Modern Takeaway

The story of the fisherman and the Djinn offers a powerful lesson about humanity's relationship with nature. Just as the fisherman's initial harmony with the sea is disrupted by overextraction and greed, today's environmental crises—such as climate change, pollution, and resource depletion—serve as a wake-up call that the forces of nature, when abused, can turn against us.

Yet, the tale also highlights the possibility of reconciliation and harmony, where wisdom, respect, and action can restore balance. In our modern world, we must recognise that nature is not our enemy, but a partner whose power we must channel with care. By making conscious choices to reduce emissions, protect ecosystems, and live sustainably, we can forge a path toward a more balanced, harmonious future, where the storm can be calmed, and coexistence is possible.

The storm does not seek vengeance; it only mirrors the imbalance we create.

Frozen Realms and Their Silent Warnings

The Arctic is not just a land of ice and snow—it is a realm of whispers. Every creaking glacier, every howling wind, and every shifting current carries an ancient story. For centuries, the indigenous peoples of the North and the mythologies of Scandinavia have spoken of beings that ruled these frozen frontiers, their presence shaping the balance of winter's dominion. But what happens when these guardians fade?

Sedna: Keeper of the Oceans' Balance

The Inuit—the indigenous people of northern Canada, Greenland and Alaska—tell of Sedna, a once-mortal woman who became the ruler of the sea. Betrayed and cast into the icy depths, she transformed into a powerful deity whose hair, when tangled, disrupted marine life. Only the shamans, through ritual and reverence, could soothe her wrath and restore balance to the ocean.

Her legend is not just a tale—it is a reflection of reality. Today, the oceans warm, ice shelves collapse, and marine ecosystems shift unpredictably. Overfishing, pollution, and rising temperatures act as the comb that is never used—leaving Sedna's hair tangled in distress. The once-reliable patterns of fish migration have become erratic, a clear warning that we have ignored the pact of respect that the ancients upheld.

Frost Giants and Retreating Glaciers

Scandinavian folklore speaks of the *Jotnar*—immense frost giants who embodied the raw power of winter. These beings, dwelling in Jotunheim, their home, ruled the icy landscapes with an unrelenting force. They were feared, yet they were necessary. Their presence ensured the land remained frozen, their breath sustaining the winter's chill.

But today, the Jotnar are retreating. Not in legend, but in reality. The great ice sheets of Greenland and Antarctica are receding, their ancient fortresses crumbling into the sea. Permafrost, once a permanent feature, is melting, releasing methane into the air—an invisible giant waking from its slumber, ready to accelerate the chaos. The giants of ice are not just vanishing; they are upsetting a balance that has existed for millennia.

Ice Sprites: Fragility in a Changing World

Among these towering legends exist the lesser-known spirits—the ice sprites. Stories tell of these ephemeral beings, delicate and shimmering, appearing only under the right conditions, dancing in the frost-laden air. Unlike the giants, they do not roar. They flicker, they gleam, and they vanish at the slightest disturbance.

These sprites are the silent echoes of the Arctic's fragility. They are the first to disappear when warmth invades the cold. Their real-world counterparts—snowflakes forming at just the right temperature, delicate ice crystals that coat the land, the seasonal layers that protect Arctic creatures—are melting before their time. The disappearance of these small, intricate formations is an early signal of a system on the brink.

Modern Reckoning: Are We Listening?

These myths are not just stories; they are warnings encoded in folklore. The people who first told these tales understood something we have forgotten: respect the balance, or face the consequences. The vanishing of Sedna's order, the retreat of the frost giants, and the fading of the ice sprites are not just metaphorical losses.

They are real, measurable, and unfolding before our eyes.

What the ancients feared was not just the wrath of their gods, but the collapse of the delicate equilibrium that allowed life to thrive. Today, we do not need shamans to interpret the signs—we have scientists, satellite data, and environmental reports. But the message remains the same.

Hope is not lost. Like the aurora that shimmers across Arctic skies, fleeting yet persistent, there is still a chance to restore balance. But the question remains: will we listen to the whispers before they are silenced?

> *The frozen whispers of the Earth warn us—respect the balance, or be swallowed by the storm.*

Aranya, the Golden Banyan Tree and the Sacred Forest

In Bharat, near present-day Lucknow, stood the sacred forest of Naimisharanya. Mentioned in both the Ramayana and the Mahabharata, it was a place of immense natural and spiritual wealth. At its centre stood a majestic Golden Banyan Tree, said to be a divine creation of the gods. Its roots extended deep into the Earth, binding life forms in a delicate equilibrium. It was believed that as long as the tree thrived, nature would remain in balance. The people revered it, understanding that their well-being was tied to the health of their environment. Under its golden canopy, wisdom flourished, rivers flowed clean, and Bharat prospered in harmony with nature.

Decline of Aranya

But time eroded reverence. As centuries passed, progress replaced wisdom, and greed overshadowed respect for nature. Aranya, once a paradise, began to wither. The Golden Banyan Tree, sensing this imbalance, lost its radiance.

Seeing the decline, the gods intervened. They summoned Vishravas, a sage known for his deep understanding of both nature and human behaviour—perhaps the sustainability consultant of his time. When Vishravas arrived, he was heartbroken. The land was barren, rivers polluted, and the once-mighty Banyan Tree stood dull and lifeless. He knew that mere words would not be enough. He needed a plan that would heal both the forest and people's hearts.

Revelation of Dharmic Exchange

One night, as he meditated under the Banyan Tree, Vishravas had a vision.

The Banyan Tree spoke to him.

> *Balance can only be restored through Dharmic Exchange—what is taken must be returned.*

The path forward was clear. Vishravas taught the people this forgotten principle:

- For every tree felled, two must be planted.
- The soil must be nourished, not exploited.
- Energy must be harnessed from nature, not against it.

Slowly, the people changed. They restored the land, cared for the rivers, and gave back more than they took. The Banyan Tree glowed once more, its golden light returning. The rivers ran clear, the air turned crisp, and prosperity found its way back to Bharat.

Whisper of the Earth

Generations later, the tale of Vishravas remains a cornerstone of Bharat's wisdom. Incidentally, Vishravas was the father of Ravana—a reminder that wisdom and destruction are often born of the same lineage. The principle of Dharmic Exchange is now taught in schools, practised in communities, and revered in temples.

But unlike past generations, who needed only one reminder, modern civilisation has forgotten the lesson many times over. Yet, as we stand at the brink of climate collapse, the whispers of Aranya grow louder. The story calls upon us once more—to honour the ancient wisdom of sustainability and act with foresight, ensuring that Earth remains a thriving home for all life.

Modern Takeaway: Dharmic Exchange in Today's World

The Dharmic Exchange is more than an ancient principle; it is the foundation of true sustainability.

- **Sustainable stewardship:** Every resource taken from nature must be replenished. Reforestation, organic farming, and renewable energy must replace exploitation.
- **Interconnectedness:** Our survival depends on the well-being of ecosystems. Protecting forests, rivers, and wildlife is not charity—it is self-preservation.
- **Balanced progress:** Development must not come at the cost of destruction. True progress is a delicate dance between economy and ecology.
- **A call to action:** Climate change is not just a crisis; it is a failure to listen. The lesson of Aranya teaches us that if we continue to ignore nature's whispers, the balance may be lost forever.

Whisper of the Earth: Remember this Punchline

Give back more than you take—only then will the Earth whisper in harmony.

Collective Whispers of the Earth

Through these tales—woven from myth, history, and lived experiences—the Earth speaks with a unified voice. Across continents and civilisations, from Rome's smoke-choked skies to the frozen landscapes of the Arctic, from the lush rice paddies of Indonesia to the vast deserts of Arabia, a singular truth emerges: humanity and the Earth are bound by an unbreakable thread, and our survival depends on recognising this connection.

Essence of Harmony

In stories like 'Lord Ganesha: The Keeper of Harmony' and 'Kami: The Spirit of Nature,' we are reminded that progress must be guided by balance. Ganesha, the remover of obstacles, teaches us that wisdom, patience, and respect for nature's rhythms are the true keys to prosperity. These narratives urge us to seek solutions that do not disrupt but rather enhance the natural world. In an age of unchecked consumption, they offer a simple yet profound reminder: to move forward, we must first learn to walk in step with nature.

Burden of Responsibility

'The Tale of Yu the Great' and 'Yudhishthira's Wisdom' carry lessons of leadership and accountability. Whether taming a mighty river or upholding Dharma in the face of adversity, these stories remind us that true leadership is not about dominion over nature but stewardship of it. They ask us to measure our decisions not just by immediate gain but by their impact on future generations.

Strength in Resilience and Adaptation

From the frozen guardians of the Arctic in 'Stories from the Frozen Realm' to the cunning resourcefulness of Anansi in 'Anansi and the Weight of Wisdom,' we see resilience woven into the fabric of human survival. These stories reflect our capacity to endure, adapt, and innovate—provided we respect the limits of the natural world.

Power of Unity and Compassion

'The Legend of the Bamboo Cutter' and 'Santa Claus and the Spirit of a Greener Christmas' emphasise the transformative power of kindness. They remind us that caring for each other and the planet is not an act of charity but of necessity. True progress is not measured in individual success but in collective well-being, in communities that thrive because they nurture both people and the Earth.

The Interwoven Web of Life

In 'Dewi Sri: The Guardian of Rice' and 'The Story of the Earth Divers,' the cyclical nature of existence unfolds. The planting, harvesting, and reaping of rice mirror the rhythms of life itself, urging us to safeguard the fragile balance that sustains us. These stories whisper an essential truth—when we harm the land, we harm ourselves. When we restore it, we restore our own future.

Whisper of Wisdom

From the depths of Maya cosmology in 'Maya Echoes' to the crossroads of culture and commerce in 'Silk Road Shadows,' these stories carry the whisper of wisdom. They tell us that knowledge is not merely a tool for human advancement but a bridge to deeper understanding. To navigate the future, we must learn from the past—not only its triumphs but also its warnings.

Universal Message

Across these narratives—whether through Sedna's icy depths, the smoke of ancient Rome, or the sacred lands of Dewi Sri—the Earth whispers a profound message:

Tread lightly. Listen deeply. Act wisely.

These stories are both cautionary tales and beacons of hope. They reveal humanity's dual nature—our capacity for destruction and our potential for redemption. They call on us to embrace the humility of the Earth Divers, the perseverance of Yu, and the compassion of Santa Claus.

As these whispers converge, they form a powerful call to action. Every choice we make—whether to conserve water, reduce waste, or nurture the soil—sends ripples across time. These stories do not simply inspire us to dream of a sustainable world; they ask us to build it, one action, one choice, one whisper at a time.

Whisper of the Earth: Sacred Chants, Sustainability, and Governance

(a) Introduction: Eternal Dance of Progress and Preservation

Humanity stands at a crossroads, where the pursuit of progress and the need for preservation engage in a delicate waltz. Our technological marvels and economic advancements have propelled civilisations to great heights, yet beneath the towering skyscrapers and digital landscapes, an ancient whisper lingers:

How do we safeguard the natural world from unchecked ambition?
How do we progress without losing the essence of what sustains us?

While modern science, policy, and innovation offer solutions, ancient traditions hold wisdom that transcends time.

The *Vishnu Sahasranamam* and *Lalitha Sahasranamam*—hymns recited for centuries—contain truths that subtly reveal the principles of balance, sustainability, and governance. These are not mere incantations of devotion but cosmic blueprints for harmonious existence.

The thousand names of Mahavishnu and Lalitha Devi illustrate the eternal need for Dharma—the righteous path that ensures equilibrium between human ambition and nature's preservation. These divine manifestations are not just celestial figures but metaphors for sustainable governance, urging us to walk with awareness, govern with wisdom, and recognise that sustainability is not a modern construct but an ancient imperative.

The Earth does not demand; it whispers. It speaks through the rhythm of the tides, the rustling leaves, and the shifting seasons. To understand sustainability, we must listen—not just with our ears, but with our souls.

(b) Vishnu Sahasranamam: Ten Names, Ten Lessons in Sustainability and Governance

The *Vishnu Sahasranamam*, a sacred hymn of the thousand names of Lord Vishnu, was first revealed in the Mahabharata, specifically in the *Anushasana Parva*. The great war of Kurukshetra had ended, and the battlefield lay silent after days of relentless bloodshed. The Pandavas had emerged victorious, but their triumph was overshadowed by grief, loss, and uncertainty about the future of governance.

Amidst this atmosphere of both devastation and new beginnings, Yudhishthira, the eldest Pandava, sought guidance on the principles of Dharma, righteous rule, and the path to inner and societal peace. It was in this profound moment that Bhishma, the grandsire of the Kuru dynasty, lying on a bed of arrows, imparted the *Vishnu Sahasranamam* to Yudhishthira.

Bhishma, despite being mortally wounded, remained steadfast in his devotion and wisdom. As he awaited the auspicious moment of his passing, he used his final days to offer teachings on governance, duty, and the eternal principles of righteousness.

At Yudhishthira's request for the supreme path to virtue and liberation, Bhishma recited the thousand names of Vishnu, declaring that remembering and chanting these names would bring peace, protection, and ultimate fulfilment. The setting was one of solemnity and divine grace. The atmosphere was charged with both the weight of past actions and the hope for a righteous future.

The *Vishnu Sahasranamam* was not just a hymn of devotion; it was an eternal guide, a reminder that true governance and sustainability must align with the cosmic order—Dharma. It symbolised balance, preservation, and the understanding that leadership, when rooted in ethical foundations, can restore harmony to a fractured world.

Each of Vishnu's thousand names is a doorway to understanding life and governance. Here are ten names deeply relevant to Environmental, Social, and Governance (ESG) leadership.

1. **Vishnu: The Protector of the Universe**

   As Vishnu preserves cosmic order, governance must ensure responsible resource management, nurturing industries while safeguarding the planet for future generations.

2. **Trivikrama: The One Who Measures the Universes**

   Trivikrama's legendary strides symbolise measured progress. True sustainability ensures economic growth aligns with environmental responsibility and social equity.

3. **Narayana: The Refuge of All**

   Narayana represents universal shelter. Sustainable development must create a safe, inclusive world, ensuring climate resilience and environmental conservation for all.

4. **Adhokshaja: The One Beyond Material Perception**

   True success is not measured by material excess but by long-term well-being. Sustainable governance must prioritise enduring prosperity over immediate gains.

5. **Purushottama: The Supreme Being**

   ESG leadership must embody the highest ethical standards, ensuring responsible decision-making that benefits both present and future generations.

6. **Vasudeva: The Source of All Good**
   Governance must be rooted in integrity and fairness, recognising that responsible policies create a just and sustainable world.
7. **Sahasranama: The One with a Thousand Names**
   A thousand names for a single divinity reflect interconnectedness. Sustainability requires a holistic vision where every decision acknowledges its societal and environmental impact.
8. **Ananta: The Infinite**
   Ananta signifies boundless existence. Resources, when managed wisely, regenerate endlessly—mirroring circular economies and renewable energy principles.
9. **Bhava: The Source of All Existence**
   Sustainability thrives on interconnectedness. Recognising the interdependence of economies, communities, and ecosystems is fundamental to responsible governance.
10. **Yogeshwara: The Lord of Yoga**
    Yoga represents unity and balance. Sustainability demands collaboration between businesses, governments, and communities to create policies that ensure harmony and equity.

(c) Dharmic Revelation: The Wisdom of Lalitha Sahasranamam

The *Lalitha Sahasranamam*, a sacred hymn extolling the thousand names of the Divine Mother, was first revealed in the *Brahmanda Purana* as part of the *Lalitopakhyana*. Unlike the *Vishnu Sahasranamam*, which emerged on the battlefield as Bhishma's final teaching, the *Lalitha Sahasranamam* was unveiled in a celestial setting, resonating with the energy of divine feminine power, creation, and cosmic harmony.

The context of its revelation was a great cosmic battle between the forces of light and darkness. The demon Bhandasura, born from the ashes of Kamadeva (the god of love, incinerated by Shiva's third eye), had become a formidable force, threatening the stability of the universe. As the *devas* and sages sought refuge, the supreme Goddess Lalitha Tripura Sundari manifested in all her radiance, seated on a celestial chariot known as the Sri Chakra, prepared to restore balance.

It was in this divine moment that the sage Agastya, an enlightened being and a key figure in many Hindu scriptures, received the *Lalitha Sahasranamam* from Hayagriva, an incarnation of Vishnu with the head of a horse, revered as the bestower of divine wisdom. Hayagriva narrated this hymn as the ultimate source of spiritual elevation, revealing that chanting the thousand names of Lalitha Devi grants wisdom, prosperity, and liberation. Unlike the battlefield solemnity of the *Vishnu Sahasranamam*, the *Lalitha Sahasranamam* was revealed in an atmosphere charged with divine grace, where cosmic forces were aligning for restoration and renewal. It emphasised *Shakti*—the dynamic energy of creation and sustenance—and its role in governance, balance, and the nurturing aspects of sustainability.

The hymn was not just a devotional recitation but a blueprint for harmonising progress and preservation, where the Earth itself was seen as an extension of the goddess, a living entity deserving of respect and care.

Lalitha Devi, the divine feminine force, embodies transformation and renewal—qualities central to sustainability and governance. Her names resonate deeply with the balance between progress and preservation.

1. **Shiva-Shakti-Aikya-Roopini: Divine Union**
   True sustainability fuses progress and preservation. Industrial growth must coexist with ecological conservation, ensuring prosperity without harming nature.
2. **Vishvagarbha: Womb of the Universe**
   Just as Lalitha nurtures all existence, the Earth must be safeguarded as the womb of all life, where resources are protected, not exploited.
3. **Panchabhoota-Mani-Mani: Embodiment of the Five Elements**
   The five elements—Earth, Water, Fire, Air, and Space—are fundamental to life. Respecting them is at the heart of sustainable governance.
4. **Jnana-Vignana-Rupini: Embodiment of Knowledge and Wisdom**
   Sustainability is not just science but wisdom in action. Theoretical understanding must lead to decisive, ethical implementation.
5. **Dharmadhara: Upholder of Dharma**
   ESG principles mirror Dharma—righteous governance that harmonises people, the planet, and prosperity.

(d) Whisper of the Earth

The Earth does not scream; it whispers. It speaks in the quiet rustling of leaves, the steady rhythm of tides, and the silent resilience of the mountains.

It reminds us that we are caretakers, not owners. It calls upon us to listen—to Vishnu's preservation, to Lalitha's nurturing, to the ancient wisdom that teaches balance.

Sustainability is not merely a policy or a trend. It is a sacred duty, a cosmic responsibility to govern wisely, to progress without destruction, and to honour the delicate dance between humanity and nature.

(e) Timeless Wisdom–A Punchline

Progress without wisdom leads to destruction. Preservation without purpose leads to stagnation. But when progress and preservation walk together, guided by Dharma, the Earth thrives, and humanity flourishes.

In whispers soft, the Earth imparts,
A call to minds and tender hearts.
From frozen realms to fields of grain,
Harmony's tune and wisdom's refrain.
Balance and care, the tales implore,
To tread with grace on nature's floor.
A bond eternal, fragile yet deep,
The Earth's embrace, for us to keep.

Inspiring Tales of Environmental Stewards

Having witnessed the effect of emissions and their categorisation, in this section let's take an inspirational journey where ordinary people achieve extraordinary feats in environmental stewardship. In the upcoming pages, we will discover the stories of seven modern-day heroes from around the world who have dedicated their lives to protecting and preserving our planet.

These aren't tales of celebrities or famous environmentalists. Instead, they are the narratives of common people, just like you and me, who saw a need in their communities and took it upon themselves to create change. These people come from diverse backgrounds with work experience on varied aspects of sustainability from reforestation and water conservation to renewable energy and sustainable agriculture.

Each story is unique, highlighting the incredible impact one person can have, regardless of their circumstances. We will learn about Jadav Payeng, known as the 'Forest Man of India,' who single-handedly transformed a barren sandbar into a thriving forest. Arun Krishnamurthy, has been restoring lakes in India and Priti Patkar is rehabilitating vulnerable communities through sustainable initiatives.

From Asia to Africa, from America to Europe, these heroes have made significant strides in environmental conservation and continue to inspire us with their dedication and resilience. As you read through each story, I hope you find not just inspiration but also actionable insights that you can bring into your own life and community.

Remember, the power to create change lies within each one of us. So, let's see what these modern-day heroes have done and celebrate their work and draw strength from their journeys as we take our first step towards a more sustainable and harmonious world.

Jadav Payeng: The Forest Man of India

Jadav Payeng, known as the 'Forest Man of India,' is a living testament to the power of perseverance and ecological restoration. Born in 1963 in Assam's Brahmaputra basin, he witnessed the devastation of a catastrophic flood in 1979, which displaced communities and ravaged the land. At just 16, he embarked on an extraordinary mission—to transform the barren sandbar of Majuli into a thriving forest.

Undeterred by scepticism and adversity, Payeng began by planting bamboo, creating the foundation for a resilient ecosystem. Over the decades, he expanded his efforts, introducing native species like neem, arjun, and cottonwood.

His dedication bore fruit as the barren land blossomed into a lush 1,360-acre forest, now home to elephants, tigers, rhinos, and countless other species.

Despite threats from the illegal logging mafia and encroachers, Payeng remained the forest's fiercest guardian, fighting to protect what he had nurtured. His efforts stabilised the riverbanks, mitigated soil erosion, and provided sustainable livelihoods to the local community.

In 2012, the Assam government honoured him with the title of the 'Forest Man of India,' and his story continues to inspire global conservation efforts. Through talks and workshops, he urges future generations to embrace environmental stewardship.

Moral: A single seed of determination can grow into a forest of change.

Arun Krishnamurthy: The Lake Saviour of India

Arun Krishnamurthy, founder of the Environmentalist Foundation of India (EFI), left a promising career at Google to restore India's dying lakes, bringing back biodiversity and securing water for future generations. His first success, Kilkattalai Lake in Chennai, became a model for restoring 93 lakes across India, combining community involvement, scientific expertise, and policy advocacy. Overcoming bureaucratic hurdles, pollution threats, and financial constraints, his efforts have transformed ecosystems, influenced urban planning, and inspired a nationwide movement. His story proves that one determined individual can spark lasting environmental change.

Moral: One ripple of action can restore an ocean of hope.

Priti Patkar: Breaking the Cycle of Exploitation and Poverty

Priti Patkar, through her organisation Prerana, has devoted her life to rehabilitating children of sex workers, ensuring they break free from the cycle of exploitation and poverty. Growing up in a family of social workers, she developed a strong sense of justice, which led her to work in red-light areas during her time at the Tata Institute of Social Sciences. Witnessing the vulnerability of children born into such environments, she established Prerana in 1986 with the mission of providing these children with safety, education, and empowerment.

Prerana's holistic approach goes beyond rescue—it includes night shelters that shield children from abuse, educational programmes that prepare them for mainstream schooling, and vocational training that helps them build independent futures. Healthcare initiatives address their physical and psychological well-being, while legal aid ensures their rights are protected. Despite immense challenges, including social stigma, resistance from brothel owners, and financial constraints, Priti's perseverance has enabled thousands of children to reclaim their dignity and build meaningful lives.

Her efforts have earned her widespread recognition, including the Ashoka Fellowship and the Nari Shakti Puraskar, India's highest civilian honour for women. More importantly, the impact of her work is seen in the lives of those who, once trapped in exploitation, are now educated, employed, and advocating for change. As she continues to expand Prerana's reach, her vision is to integrate rehabilitation with broader community development, ensuring a sustainable and just future.

Moral: When compassion meets action, even the most marginalised can rise and thrive.

Wangari Maathai: The Green Belt Movement and Beyond

Wangari Maathai, a Kenyan social activist and founder of the Green Belt Movement (GBM). She empowered women to plant over 50 million trees, addressing deforestation, environmental crises, and sustainable livelihoods. Inspired by her deep connection to nature and the struggles of rural communities, she mobilised women to restore landscapes through tree planting, organic farming, and water conservation. GBM's efforts not only revitalised ecosystems but also improved food security, income, and environmental awareness.

Beyond conservation, Maathai championed women's rights, democratic governance, and social justice, using GBM as a platform for broader change.

Moral: One tree planted with purpose can grow into a movement that transforms both lives and landscapes.

Anote Tong: Championing Climate Change Awareness for Kiribati

Anote Tong, former President of Kiribati (2003–2016), emerged as a leading voice in the global fight against climate change, highlighting the existential threat faced by his low-lying island nation. Born in 1952, he witnessed firsthand the rising seas and their impact on his homeland. Educated in New Zealand, he adopted a pragmatic, evidence-based approach to governance.

As President, Tong confronted Kiribati's vulnerability—33 atolls and reef islands at risk from storm surges, coastal erosion, and saltwater intrusion. He saw climate change as a national security crisis and took urgent action, constructing seawalls, rehabilitating mangroves and coral reefs, and improving rainwater harvesting to secure freshwater resources. Understanding that adaptation alone was not enough, he explored long-term resilience strategies, including the controversial yet necessary step of purchasing land in Fiji as a potential refuge for displaced citizens.

On the global stage, Tong amplified the voice of vulnerable nations. His UN speech, 'We are the face of climate change,' was a stark call to action, urging major emitters to recognise their moral responsibility. He played a key role in strengthening the Alliance of Small Island States (AOSIS), advocating for ambitious climate commitments from developed nations. He championed the principle of 'common but differentiated responsibilities' to push for stronger global policies and financial support for adaptation.

Despite political and resource constraints, Tong's leadership kept climate change at the forefront of international discussions. His legacy is a testament to the power of advocacy and strategic action in the face of an impending crisis.

Moral: Leadership is not just about governing—it's about securing a future for those yet to come.

Yacouba Sawadogo: The Man Who Stopped the Desert

Yacouba Sawadogo, a farmer from Burkina Faso, gained global recognition for his pioneering work in reversing desertification in the Sahel. Known as the 'Man Who Stopped the Desert,' he revived and refined traditional farming techniques, transforming barren land into fertile fields and strengthening local agriculture. His story is one of resilience, ingenuity, and an unwavering commitment to restoring the environment.

Born in the late 1940s in Gourga, northern Burkina Faso, Sawadogo grew up witnessing the harsh realities of desertification—crop failures, food shortages, and an ever-expanding Sahara. With limited formal education but deep ecological knowledge, he sought solutions when severe droughts in the 1970s and 1980s devastated local farming.

Sawadogo revived the *Zai* technique—small pits dug

into the soil to trap rainwater and organic matter, improving seed germination. He innovated by making the pits larger and enriching them with compost, significantly boosting their effectiveness. He also used 'contour stone bunds,' where stones placed along field contours slowed runoff, reducing erosion and enhancing water infiltration.

Initially met with scepticism, his methods gained traction as his once-barren fields flourished. His success drew the attention of researchers and development organisations, leading to wider adoption of his techniques across the Sahel. He established demonstration farms to train others, scaling up his impact.

The results were transformative—thousands of hectares of degraded land were restored, food security improved, and farming communities became more resilient to climate change. In 2010, he received the Right Livelihood Award for demonstrating that farmers could regenerate their land using indigenous knowledge without costly external inputs. His work was also featured in the documentary, *The Man Who Stopped the Desert*, inspiring global efforts in sustainable land management.

Sawadogo's journey underscores the power of traditional wisdom, local action, and persistence. His approach continues to influence global strategies against desertification, proving that solutions to some of the world's most pressing environmental challenges lie within the knowledge of those closest to the land.

Moral: Indigenous knowledge is powerful enough to stop the desert.

Marinel Ubaldo: Champion of Climate Justice and Resilient Communities

Marinel Ubaldo, a climate activist from the Philippines, has emerged as a powerful voice for climate justice, shaped by her harrowing experience as a survivor of Typhoon Haiyan. Born in a coastal village in Eastern Visayas, she witnessed firsthand the increasing intensity of extreme weather events driven by climate change. When Haiyan struck in 2013, it devastated her community, claiming thousands of lives and displacing millions. This catastrophe ignited her resolve to fight for those most vulnerable to climate change.

Understanding the socio-economic struggles of her people, Ubaldo committed herself to raising awareness and pushing for systemic change. She highlights the stark injustice that the communities most affected by climate change contribute the least to its causes. Her advocacy extends beyond recounting the destruction—she demands action, urging governments and institutions to prioritise resilience and justice.

Ubaldo has become a key figure in the global climate awareness movement, representing marginalised communities at international forums like the United Nations Climate Change Conferences (COP). She calls for financial and technological support to help frontline communities adapt, emphasising that climate justice is not just about mitigation but survival. Her powerful testimony of surviving Haiyan has resonated worldwide, humanising the abstract issue of climate change.

Her grassroots efforts focus on equipping communities with tools to withstand climate threats. She has worked on sustainable livelihood programmes such as organic farming and eco-tourism, providing alternatives to fishing and agriculture, which are increasingly impacted by climate change. She also spearheads disaster preparedness training, ensuring communities can respond effectively to extreme weather events.

Education is central to Ubaldo's work. She believes young people must be empowered with knowledge to build resilient communities. She conducts workshops in schools and universities, teaching climate science, sustainability, and activism, inspiring the next generation of climate leaders. Recognised globally, Ubaldo has been named a Global Youth Ambassador for the Youth Climate Action Network, with her advocacy featured in major media outlets. Despite accolades, she remains grounded, focused on amplifying marginalised voices, pushing for equitable climate policies.

Her journey—from a typhoon survivor to a global advocate—is a testament to resilience and justice. As the world grapples with escalating climate crises, Ubaldo's story serves as a compelling call to action, reminding us that real change starts with those who refuse to be silenced. Her work continues to inspire a new wave of activists dedicated to safeguarding both people and the planet.

Moral: It needs just one voice to raise global awareness and fight for justice.

Your Role in the Journey

This session plans to demonstrate what you as an individual can do. There are plenty of takeaways. The aim of this session is to not just provide a high-level understanding of emissions and leave it at that.

It is like a call to action to you, the individual, who holds immense potential to effect change. Irrespective of your position in society, be it as part of a government, a city, a company, an organisation, a community, or even as an individual, your own actions can significantly influence the collective effort towards reducing emissions and promoting sustainability.

This journey is not solely the responsibility of governments and large corporations; it is a shared endeavour. Every active participation by each person counts. All it requires is you making conscious choices and adopting sustainability measures that contribute to a healthier planet.

Simple Story, Great Impact

Rolling hills and lush fields surround the small town of

Varnan. Imagine that is where you live. It is here where you were instrumental in performing a single act of kindness, that sparked a movement, that created a ripple across the world!

You might wonder, why does it have to be me?

Can a single act by me create an effect at the world level?

This is not imaginary.

It is possible. Just a single act of kindness has that potential.

Schoolteacher in Varnan

Imagine you are doing it now, as a schoolteacher in Varnan with a simple idea that laid the foundation for a sweeping transformation.

One afternoon, you decided to attend the local council meeting, where the community discussed the upcoming elections and the future of the environment surrounding Varnan. The debate was intense, with opposing views about the direction the town should take.

You have a deep understanding of the earth's fragility and seeing an opportunity, you stand up and with a steady voice urge your fellow citizens to support leaders who prioritise sustainability, urging them to vote for policies that protect Varnan's environment and create jobs. Your speech was instinctive, but not the usual run of the mill speech. It was a call to action, not just for the council but for every individual present.

That night, you aren't able to sleep and heightened adrenaline keeps you awake. You are restless; lying down is a waste of time. You think and you pace the room for a while. You switch on your laptop and share your thoughts on social media, calling upon others to raise awareness.

You write a few lines.

We're not inheriting the Earth, we're borrowing it–let's return it better than we found it.
The world won't fix itself. The change we want? We must be it now.
Tomorrow's world isn't a distant dream–it's what we create today. Don't wait.
The planet is speaking. The question is, are we listening or just scrolling?
If we don't act, the future won't be out of reach–it'll be lost.
Let's turn the tide, together.
Only by participating in our democracy can we ensure the future of our children.

An innocent message written in response to the pain caused by the actions of the sand mafia of Varnan catches the town's attention and then spreads beyond. The post goes viral.

Soon, people from all walks of life absorb this message, organising public consultations, civil groups, and environmental advocacy campaigns.

Your words sparked a movement that brought not only voters to the polls but also citizens to action, each feeling the weight of their responsibility.

Village after village, town after town, and city after city identified with the message and linked it to a particular problem in their environment and asked for accountability from their leaders. The media picked up this viral movement and amplified it further.

While the discussions continued, something changed at Varnan. The town embraced greener policies, they took an oath to put an end to the sand mafia and acted, holding the government accountable for not taking action. The town embraced greener policies; they turned their focus towards urban sustainability. Inspired by you, many began to promote public transportation, convincing the town council to introduce more buses, better cycling lanes, and safer pedestrian zones. Local green spaces flourished, with every park and vacant lot becoming a potential garden or a green roof.

You realise the importance of communities playing a vital role in promoting environmental sustainability and organise tree plantation events, setup community gardens, hold clean-up drives, establish recycling programmes, promote eco-friendly practices, promote local crafts and products, implement carpools, create a community composting programme, advocate community solar initiatives, collaborate with local schools, organise sustainable cooking classes, eco-friendly community festivals, organise workshops on home energy audits, and mindful waste management, apart from numerous other programmes.

You identify deeply with the collective action and appreciate the changes taking place around you. Support comes from unexpected quarters.

You find your calling and become a spokesperson for the national green movement. A thought can change a town; it can also change the city, state, and country. You become an evangelist of policy advocacy that prioritises environmental responsibility. You become a voice of the citizens and urge the audience to adopt a sustainable lifestyle. We must do more than recycle. We must live with the Earth in mind in every action and every decision. You speak and people hear.

Once they hear, they act.

The Playbook

You bring out a playbook for every citizen to adopt. It mentions a long list of actions every citizen can adopt in their daily life.

- Switch to LED light bulbs
- Unplug electronics when not in use
- Use energy efficient appliances (of at least three stars and above)
- Install solar panels
- Use programmable thermostats
- Install double-glazed windows for better insulation
- Use energy-efficient curtains or blinds
- Conduct home energy audits
- Install a tankless water heater
- Seal air leaks around doors and windows
- Use smart power strips
- Wash clothes in cold water
- Air-dry clothes instead of using a dryer
- Use a clothesline outside
- Insulate your attic and walls
- Replace old HVAC systems with energy-efficient models
- Use ceiling fans to reduce use of air conditioners
- Install a rainwater harvesting system
- Use a low-flow showerhead
- Install a dual-flush toilet
- Use a water-saving dishwasher
- Use reusable water bottles and avoid single-use bottled water
- Fix leaky faucets promptly
- Grow vegetables in your balcony or terrace
- Install a greywater system for home irrigation
- Use ride-sharing services or join a car-pooling programme
- Invest in an electric vehicle
- Maintain proper tire pressure
- Drive at optimal speed for fuel efficiency
- Avoid idling your car
- Use aerodynamic roof boxes on cars

- Encourage biking in your community
- Install a home EV charging station
- Have a 'no drive day' in your life
- Eat seasonal and local products
- Reduce food waste
- Compost kitchen scraps
- Think to buy second-hand items
- Host clothing swaps with friends
- Donate unused items
- Use natural cleaning products
- Choose products with minimal packaging
- Avoid fast fashion
- Make your own household products, as much as possible
- Volunteer for environmental causes
- Create a sustainable living group
- Reduce paper usage by going digital
- Repair items instead of replacing them
- Walk instead of driving, whenever possible
- Avoid single use plastic
- Choose ecofriendly furniture
- Reduce water heater temperature
- Choose organic produce
- Make plant-based meals a regular habit
- Buy in bulk to reduce packaging
- Participate in local tree planting drives
- Opt for natural personal care products
- Repeatedly wear the clothes you buy

Grassroot changes get recognised by larger corporates who take note. One of Varnan's largest employers, a manufacturing company called Harmonica Industries, picks up on the ripples in society.

The CEO Raj, recognises the potential of green practices to transform the business, while contributing to global sustainability goals. Raj implements energy management systems, installs solar panels on factory rooftops, and incentivises employees to reduce their carbon footprint. In the office, workers adopted sustainable practices. Harmonica Industries organises workshops on green technologies and you are called to be part of numerous workshops.

These actions born out of corporate responsibility inspire others to follow suit, creating a domino-effect across industries. Other companies take note and adopt eco-friendly practices. Governments, under pressure from corporates and citizens alike, pass laws that mandate the reduction of greenhouse gas emissions and incentivise renewable energy projects. But the true power of this transformation lies in the collective actions of millions. It wasn't just the leaders in boardrooms or policymakers in government that shaped the future. It was individual action that did the work.

At the community level, people continue to support local farmers' markets, develop shared spaces for community gardening, and advocate for solar energy projects. They join forces with local NGOs and rally behind eco-friendly policies, creating a network of support that spans the nation. In towns and cities, the green wave continues to spread.

Public transportation systems become the norm. Residents organise 'no car days,' as they cycle and walk through the streets, while electric vehicles (EVs) become a common sight, their emissions almost negligible. People buy in bulk, reducing packaging waste, and local grocery stores are filled with organic, plant-based foods that are both affordable and delicious.

Communities come together to teach sustainable cooking, repair old furniture, and offer low-impact solutions to daily challenges. The days of mindless consumption taper off; every purchase is done with the planet in mind.

Years pass, and the results are evident. Emissions have plummeted, urban air quality has improved, and green spaces flourish in cities that were once concrete jungles. The world begins to take notice as this small town's ripples turn into a global wave.

You are Varnan's elder, looking over the community with pride. You knew the road was long, but the journey continues. Your story becomes a blueprint for other cities and towns in numerous countries. The message is clear.

Change starts with the individual, but is only truly powerful when it spreads to communities, organisations, and governments. And the key to it all is simple. We are all in this together. The earth's whispers have been heard and the world is finally listening.

I hope you enjoyed the journey, with yourself as the protagonist.

A small action from you can create a world-changing ripple effect. Sustainability is everyone's responsibility. It is a citizen's movement, a community movement, a city/statement movement, and the government's top concern, but all that starts with you. It has an impact towards a very serious issue that the Earth is facing—climate change.

This narration by itself has plenty of start-up ideas unleashing entrepreneurial and innovative mindsets. There is a huge unmet demand in renewable energy sources, energy efficient appliances, and smart grid technologies. They all present numerous entrepreneurial avenues.

Start-ups focusing on solar energy, wind energy, and energy storage solutions are not only viable but crucial for a sustainable future. Massive untapped potential lies in EVs, green logistics practices, and creation of a smart transportation system. They offer fertile ground for new businesses. Entrepreneurs can investigate ride-sharing platforms, electric vehicle charging infrastructure, and other eco-friendly logistics solutions. A comprehensive end-to-end waste recycling programme is non-existent. There are players who cannot meet the demand requirements in this untapped domain.

Start-ups can explore recycling technologies, composting solutions, and sustainable packaging alternatives. Technologies for water-saving fixtures, rainwater harvesting, and greywater systems are essential for sustainable living. Entrepreneurs can develop products and services that help conserve and efficiently manage water resources. Another community initiative that appears to be untapped is urban gardening, community-supported agriculture, and local clean-up drives. These initiatives not only reduce emissions but also strengthen community bonds.

Starting community-focused businesses that promote sustainable living can have a profound impact. The important point is building of sustainable culture practices in our daily lives. This involves educating and raising awareness about the importance of sustainable practices.

Another area where citizens, organisations, and communities can get involved in is in becoming an advocate for green policies. This is critical. Policymakers need inputs. One can be an advocate in this area by supporting leaders and policymakers providing input on subjects related to environment protection, civil society participation initiatives, and spreading knowledge about the benefits of low carbon practices.

All this may sound boring, but they are fundamental to systemic change. National and local policies may set the framework, but public support and participation are essential for their success. Be an advocate.

It might appear daunting initially. It is all a matter of taking the first step and here are some stories to reinforce this thought.

Hanuman's Leap: Journey Across the Sea

Hanuman stood on the shores of the mighty ocean, gazing at the endless expanse of water. On the other side lay Lanka, where Sita was held captive. Ram's message to her was urgent, and it was Hanuman's task to deliver it. The ocean seemed vast, and his mission daunting. But then he remembered his purpose. Drawing strength from his devotion to Ram and the greater good of their mission, he planted his feet firmly on the ground, breathed deeply, and leaped into the sky.

Mount Mainaka: Comfort of Rest

As Hanuman soared through the air, the shimmering waters below whispered promises of ease and rest. Suddenly, a golden peak emerged from the ocean. It was Mount Mainaka, a benevolent guardian of the seas.

Rising to meet Hanuman, Mainaka spoke: 'Brave son of the wind, your journey is long and arduous. Rest upon my peak and regain your strength before continuing.' Hanuman paused, touched by the mountain's kindness. But his heart knew that time was precious. 'Your offer humbles me, oh

Mainaka,' he said, bowing, 'but my mission cannot wait. I must press on.' With a final nod of gratitude, he soared higher, leaving the mountain behind.

Hanuman's refusal to rest echoed a vital truth: the path to fulfilling one's purpose often demands resilience over comfort.

Surasa: Test of Wisdom

Further along, the sky darkened, and a fearsome figure emerged from the clouds. It was Surasa, the mother of serpents, her jaws as wide as the sky. She challenged Hanuman with a booming voice: 'None may cross the ocean without entering my mouth. It is fate's decree that I devour you!' Hanuman smiled. 'If it is fate, then so be it,' he said, shrinking his form to the size of a thumb. He darted into her gaping mouth and out again before she could react.

Surasa was astonished. 'Your cleverness honours you, son of the wind. Go forth and fulfil your mission.'

Hanuman's encounter with Surasa revealed the power of wit and agility in overcoming obstacles. He had not fought brute force with force but had found a way around it.

Simhika: Shadow of Doubt

As Hanuman continued, he felt a strange pull from below. Glancing down, he saw a shadowy creature rising from the depths of the sea. It was Simhika, a demon who trapped her prey by seizing their shadow. She lunged at him, dragging him downward, her claws like chains of doubt and despair. But Hanuman was not one to falter. With a roar, he expanded his form, breaking free of her grasp, and struck her down with a single blow.

Simhika's defeat was a testament to the strength of

unwavering resolve. Doubts and fears could not tether him; his spirit was too vast.

Reaching Lanka

At last, the shores of Lanka came into view. Hanuman landed lightly and took a moment to survey the land. The air was thick with the scent of challenge, but his heart swelled with determination. Finding Sita in the Ashoka grove, he delivered Ram's message, rekindling hope within her.

Before leaving, Hanuman decided to sow the seeds of disruption against Ravana's oppressive rule. He set the city ablaze, not with destruction but with a spark of Ram's justice, igniting a chain of events that would lead to Lanka's liberation.

Lessons from Hanuman's Leap

Hanuman's journey was not just a physical crossing of the

sea. It was a tale of courage, intellect, and persistence. Every obstacle he overcame rippled outward, inspiring others and moving the mission forward. In the same way, each of us faces the oceans we need to cross and the challenges we need to conquer.

With purpose in our hearts and wisdom in our actions, we can create ripples that extend far beyond ourselves, transforming the world for the better.

Arjuna's Quest for the Pashupatastra

The Mahabharata war loomed on the horizon, and the Pandavas needed divine weapons to combat the mighty Kauravas. Krishna advised Arjuna to seek the Pashupatastra, the celestial weapon of Lord Shiva. However, acquiring this divine boon required immense effort, unwavering devotion, and the courage to face trials.

The Ascent to Kailasa

Arjuna embarked on his journey to Mount Kailasa, the abode of Shiva, determined to earn the lord's favour. The path was treacherous, cutting through dense forests and snowy peaks. Every step tested his resolve, as natural elements and inner doubts tried to deter him. Yet, his unwavering devotion to Dharma and Krishna's guidance fuelled his will. Through meditation and self-discipline, Arjuna immersed himself in prayer, radiating pure intent to serve the greater good.

The Test of Humility: The Boar Incident

One day, as Arjuna meditated, a fierce boar appeared, disturbing his focus. Without hesitation, he drew his bow and aimed. At the same moment, another arrow flew, striking the boar simultaneously. Arjuna turned to see a forest hunter standing before him, also claiming the kill. The hunter was none other than Shiva in disguise.

A duel ensued, showcasing Arjuna's skill, courage, and determination. Despite his best efforts, Arjuna found himself overpowered. At last, recognising the divine nature of his opponent, he surrendered, placing his weapons at the hunter's feet.

Shiva, pleased by Arjuna's humility and dedication, revealed himself and granted him the Pashupatastra.

Lessons from Arjuna's Quest

- **Perseverance against all odds**: The journey to Kailasa symbolises the uphill battles we face in life, demanding resilience and patience.
- **Focus amid distraction**: The boar incident highlights the importance of staying true to one's goals, even when

confronted by sudden challenges.

- **Humility as strength**: Arjuna's ability to recognise a force greater than himself and bow to it, reflects the wisdom in humility, leading to growth and success.

Relevance to Modern Life

Arjuna's journey mirrors the challenges in tackling complex issues like sustainability and climate change. Each step in the right direction requires overcoming physical, mental, and emotional obstacles.

His focus on Dharma can inspire us to align our actions with a higher purpose, ensuring that our struggles create a ripple effect of positive change.

The Merchant and the Ogre (From Jataka Tales)

Once upon a time, a group of merchants decided to journey across a stretch of wilderness to trade their goods. The path was perilous, for the wilderness was not only arid and barren but also home to a ferocious ogre who delighted in tricking and devouring travellers. Despite the dangers, the merchants prepared for their journey, determined to reach their destination.

The Two Leaders

The group was divided into two caravans, each led by a merchant. The first leader, arrogant and overconfident, believed the journey would be straightforward and dismissed the warnings about the ogre. The second leader, wise and cautious, listened intently to the advice given by experienced travellers. The wise leader decided to prepare meticulously. He instructed his men to carry plenty of water, food, and other essentials, and exhorted them to remain vigilant. The arrogant leader, however, scoffed at such precautions, taking only the bare minimum, thinking it would lighten the load and speed up their journey.

The Ogre's Deception

As the two caravans progressed, the ogre devised a cunning plan to trap them. Disguising himself as a fellow traveller, he approached the first caravan and announced, 'Turn back! There's no water ahead. I barely escaped with my life. You must take another path, or you'll perish!'

The arrogant leader, already low on water and supplies, panicked. Trusting the stranger without question, he turned his caravan around and walked right into the ogre's trap.

None survived. When the ogre approached the second caravan with the same lie, the wise leader was not fooled. He reasoned, 'Why should I trust a stranger? We have prepared well and are capable of completing this journey.' He instructed his men to press on, despite the ogre's warnings.

Overcoming the Final Trial

As the caravan moved forward, the heat grew unbearable, and the men began to doubt their leader's decision. They complained of thirst and fatigue, but the wise leader encouraged them, saying, 'We have come too far to turn back now. Trust our preparations and the path ahead.' Finally, they spotted an oasis in the distance. As they approached, the ogre, furious at their persistence, revealed his true form and tried to block their way. But the caravan was well-prepared. Armed with tools and unity, they fought off the ogre, reached the oasis, and completed their journey safely.

Lessons from the Merchant and the Ogre

- **Preparedness and planning:** The wise merchant's careful preparation saved his caravan, showing the importance of forethought in the face of challenges.
- **Critical thinking:** The ability to question and reason beyond appearances is key to avoiding deception.
- **Courage and unity:** Facing trials requires courage, but collective effort and trust in leadership can overcome even the greatest threats.

Relevance to Modern Life

This story is a timeless reminder of how preparation, rational thinking, and perseverance can help us navigate life's challenges. In a modern context, it can inspire individuals and organisations to face environmental crises or social upheavals with resourcefulness and unity.

Goddess Durga and the Battle Against Mahishasura

In ancient times, the universe faced a dire threat from the demon king, Mahishasura. A powerful asura, Mahishasura had acquired a boon that no god or man could kill him,

making him nearly invincible. Empowered by his perceived immortality, he unleashed terror across the three realms—Heaven, Earth, and the Underworld. His tyranny disrupted natural balance, causing suffering and despair among gods and mortals alike. The gods, unable to defeat him despite their might, gathered to seek a solution.

Realising that Mahishasura's boon excluded women, they pooled their divine energies to create a warrior like no other—a being of immense power and unmatched valour. Thus, Goddess Durga was born, radiant like a thousand suns and embodying the collective strength of the universe.

Preparation for Battle

Armed with divine weapons bestowed by the gods—Shiva's trident, Vishnu's discus, Indra's thunderbolt, and more—Durga mounted her lion, a symbol of courage and strength. Before engaging Mahishasura, she meditated deeply, drawing energy from her inner reservoir of resilience and divine purpose.

Her resolve was unwavering. She understood the magnitude of the task before her—not only to vanquish Mahishasura but to restore balance and hope to a world on the brink of despair.

Three-Day Battle

When Mahishasura saw Durga, he laughed mockingly, dismissing her as an insignificant woman. But Durga, calm and composed, challenged him to a duel. The battle raged for three days and nights, a fierce clash of might and strategy. Mahishasura, a shapeshifter, tried to confuse Durga by transforming into various forms—a lion, an elephant, and finally, a raging buffalo. But Durga remained undeterred. With her sharp intellect and unyielding focus, she anticipated his every move.

The Final Confrontation

On the third day, as Mahishasura lunged toward her in his buffalo form, Durga leaped into action. With her trident, she pierced his heart, bringing an end to his reign of terror. The Earth, the skies, and the heavens echoed with the sound of victory as the universe rejoiced. Mahishasura's defeat was not merely the vanquishing of a tyrant but a symbolic victory of good over evil, courage over fear, and righteousness over arrogance.

Lessons from Durga's Triumph

- **Inner strength**: Durga's power came from within, showing us the importance of tapping into our inner resilience during challenges.
- **Unity and collaboration**: Her creation was a collective effort, reminding us of the strength in unity when facing seemingly insurmountable odds.
- **Perseverance against adversity**: Despite the length and intensity of the battle, Durga's unwavering focus led her to victory, emphasising the value of persistence.
- **Overcoming ego**: Mahishasura's arrogance blinded him to the possibility of defeat, while Durga's humility and clarity led her to triumph.

Relevance to Modern Context

Durga's story resonates deeply in the modern era, where challenges such as environmental degradation, social inequality, and political discord may feel overwhelming. Her journey teaches us to:

- Harness collective effort to address global crises, just as the gods united to create her.
- Face problems with courage and resilience, drawing upon inner and shared resources.
- Believe in the transformative power of action, even when the odds appear insurmountable.

Beneath the sky, where shadows play,
The Earth does whisper, soft and grey.
Rise like Durga, fierce and kind,
With courage bright and purpose aligned.

Cross the seas, where fears may dwell,
Like Hanuman, with stories to tell.
Unite as one, let love abide,
For in your strength, the Earth confides.

Vasudhaiva Kuṭumbakam
The World Reborn

This section discusses what the world might look like in the next thirty years.

Which prism should be used to create scenarios for the future?

Considering that even a family is a complex unit, what we are trying to discuss here are scenarios for not just a community, a city, a district, a state, or even a country—we are trying to imagine the future. And this is tough.

Some possible macro environments are: National power, geopolitics, economics, military, politics, culture, environment, development, people, jobs, technology, natural resources, human capital, innovation, healthcare, entertainment, multi-lateral bodies, and so on.

A scenario building of the complex world, considering all these macro environments would be a mind-numbing exercise.

That is the reason I am handing over this section to a very special guest who will take you through to the end.

This special guest is Earth, you know her as Bhūmi Devi; you have met her before, when we discussed the story from the *Daśāvatāra* at the start of this journey.

Imagine Bhūmi Devi as a graceful woman, a figure of beauty and wisdom gained by the millions of years that she has endured and experienced. She walks gracefully into the room with an aura that commands both awe and warmth. Her presence brings a sense of calm and the weight of ancient knowledge along with her.

She speaks to you softly, but with undeniable authority, her voice laced with the stories of the ages.

I am the totality of everything that you and all who came before you have gone through. I am the totality of all the power that exists.

When you look at me, you witness a complex dance of elements that have created not just life, but the very essence of life itself.

She starts with her eyes and with a lot of depth in every word she says:

This is my military power. It's the protective shield I offer my children, for all the abuse that you do, I stand strong protecting life. While you violate liberty, I do my job protecting life.

Her hands, delicate yet firm move as she explained:

I am your economic power. You see, I give, I create, I nurture, and allow you to thrive. When my resources are squandered, I react and make you reflect, hoping you will understand the balance between giving and taking.

Her fingers traced invisible lines in the air.

This is my political power, the laws and order that govern all who live on me. I don't interfere unless necessary, but I ensure that justice flows. My political system is as intricate as the ecosystems that flourish here. If one part suffers, the rest is affected. Unity is what I seek, balance in every corner.

The subtle influence that I exert even when you are unaware is my cultural power. I influence your hearts, minds and your beliefs. I speak through art, stories, love and joy. I am your muse. I am the fire in your soul.

You call me Bhūmi Devi, beautiful not just for my physical form but for the richness of my spirit.

Bhūmi Devi, slowly moves towards a window and continues speaking as she gazes outside. You could hear her speak as you saw her reflection in the glass.

I am your technological power, the ingenuity that you harness–that ingenuity is what I give you, from the simplest stone tool to the intricate machine is all given by me.

I am the foundation and you are the creator and I want you to keep that in mind. I am your natural resource, the heartbeat of your life. From the minerals deep within me to the winds, water, and air—all are provided by me.

These are my gifts and they are not infinite. You must remember that. I am not just a resource for you.

I am your human capital. I am the wisdom you learn from, the knowledge you gather, the growth you experience as my children. My health is reflected in yours. If I suffer, so will you.

There was a warning in that last line....

Bhūmi Devi pauses, her hands gently resting on her heart.

Dear, the totality of all your power is not force or wealth. It is balance. It is about recognising me in all forms, understanding how each part connects to the other. When you respect this, when you care for me, I will return that care in ways you cannot even imagine.

Bhūmi Devi, looks directly into your eyes now, and in her gaze, you see a reflection of your future.

I am here, and I wait for you to recognise the power that lies in harmony with me. You are my children, and I will give you everything, but remember, the totality of your power lies in the choices you make with what I offer.

With that she extends her hand reaching out to you.

As you stretch your hand, Bhūmi Devi takes your hand in hers. You feel a powerful heat within her grasp. You wonder if Bhūmi Devi has a fever. The rise in her temperature is due to the pollution caused by humans. She needs to be treated and the only treatment is to change our way of living.

Bhūmi Devi realises where your thoughts are taking you.

With a gentle smile, she whispers: *Vasudhaiva Kuṭumbakam.*

Bhūmi Devi's whisper lingers in the air, its meaning resonating within you.

The world is one family.

This truth has been spoken across millennia, yet its essence is often forgotten in the haze of modern life. Let me tell you a story.

Snehapuri and the lesson of Vasudhaiva Kuṭumbakam

Snehapuri, a small village on the banks of the Narmada River, was a thriving community bound together by the principle of *Vasudhaiva Kuṭumbakam*. Here, people lived not just in harmony with each other, but also with the Earth. They saw the river not as a resource, but as a mother who nurtured them with her flowing waters.

One family, the Mishras, had been the keepers of an ancient tradition passed down through countless generations. They called it the *Dhara Vrata*, the Earth Vow. Every member of the family, upon reaching adulthood, would plant and care for ten trees. These trees were not chosen randomly but carefully selected to balance the needs of the land, the river, and the people. Fruit-bearing trees for sustenance, flowering trees for pollination, and hardy species to prevent soil erosion along the riverbank.

But the Dhara Vrata was far more than about just planting trees; it was a way of life. The Mishras believed that every action, no matter how small, rippled outward, affecting not just their family, but the entire world. When they cooked, they ensured no food was wasted. When they built, they used materials that would return to the Earth without harm. And when they prayed, they thanked the rivers, the soil, the sun, and the winds as part of their family.

One year, the monsoon failed. The Narmada waters receded and the fields turned to dust. The village faced a crisis unlike any they had ever seen before. Crops failed, hunger loomed, and despair crept into the hearts of even the most resilient. It was in this time of trial that the true meaning of *Vasudhaiva Kuṭumbakam* emerged. The Mishras, along with the villagers, convened under the great Banyan tree at the heart of the village. As the village elder, Aarti Mishra, spoke.

'We are not alone in this struggle,' she said, her eyes sweeping over the gathered crowd. 'The river, the trees, the animals—they are all a part of our family. We must care for them as we care for ourselves. Let us share what we have, no matter how little, and work together to restore balance.'

With that, the villagers began their most arduous undertaking. They dug trenches to capture every drop of rain when it finally arrived. They shared seeds and worked the parched fields together, ensuring no one was left behind. The children of the village were taught to care for young saplings, planted to replace the dying trees. Even in their hunger, they left offerings of grain for the birds and animals, believing that abundance would return, only if shared.

Months passed, and their efforts bore fruit—literally and metaphorically. When the rains finally returned, they found the land ready to receive the waters. The river swelled back to life, and the fields turned green once more. The village celebrated, not just for their survival but for the bonds that had deepened among them and with the Earth. Years later, the story of Snehapuri spread far and wide. People came to learn from the villagers, inspired by their resilience and their unwavering belief in *Vasudhaiva Kuṭumbakam*. The village became known, not just for their environmental stewardship, but for their human values, compassion, resilience, and unity.

Bhūmi Devi's gaze deepened as she spoke:

'The world you inherit thirty years from now will reflect the choices you make today. "Vasudhaiva Kuṭumbakam" teaches us that every action, mentioned in the Upanishads and in the "Atharvana Veda", is a way of life.

'The family we speak of is not limited by time; it includes the unborn generations, the rivers yet to flow, the forests yet to bloom, and the lives yet to breathe.'

Bhūmi Devi paused, her voice steady but filled with an unmistakable urgency.

'The question is not whether the future will arrive—it always does. The question is: What kind of future will you welcome? Will it be one where my rivers run dry, my forests burn, and the air grows heavy with despair? Or will it be a future where harmony flourishes, where humanity understands that to harm me is to harm itself?'

Bhūmi Devi's voice softened, imparting a final lesson to her children.

'The power to shape this future rests in your hands. You must redefine wealth—not in gold and stone, but in terms of clean air, fresh water, and the resilience of your communities. You must measure success not by dominance, but by balance and the capacity to coexist.'

She turned toward the horizon, the light of the rising sun streaming through the window. '*The Earth will continue its journey, as it has for eons. Whether you walk this path with me as caretakers or adversaries is your choice. But know this—your survival, your joy, and your fulfilment are intertwined with mine. I am not just your home; I am your past, your present, and your future. As the dawn of thirty years hence approaches, let this be your guiding star: "Vasudhaiva Kuṭumbakam." The world is one family. If you honour this truth, the future will not be just something you survive, it will be something you cherish.*'

With that, Bhūmi Devi smiled—a smile tinged with hope and resilience. She stepped back into the light, her figure dissolving into the golden glow of the horizon. For a moment, the room was silent, as if the Earth, herself, had paused to listen.

And then you realised: the future she spoke of was not distant. It had already begun, here and now, with the choices you were about to make.

Vishnu's Giant Strides

Bali was not merely a king, but a ruler whose heart pulsed with the rhythm of his people's joys and sorrows. The farmers in his fields sang songs of gratitude; the traders spoke of his fairness. Yet, Bali was human, and in every human heart, there resides a spark of ambition.

One starlit night, standing on the terrace of his palace, Bali gazed at the heavens. The stars seemed so close, almost within reach. 'Why must I stop here?' he whispered. 'Why not the skies too? Why not the realms of the devas?'

His trusted advisor, Prahlad, his wise grandfather, and a devotee of Vishnu, often cautioned him. 'Bali, ambition is a double-edged sword. In seeking more, do not forget what you already hold—a kingdom in harmony.'

But free will is both a blessing and a burden. Bali chose to march forward, his desire pulling him like a moth toward the flame of the heavens. He planned a great *yajna* to expand his powers—and to even take control of the heavens.

The morning of the grand *yajna* arrived, the air thick with the scent of sandalwood and the hum of Vedic chants. Bali's court was a mosaic of nobility, sages, and commoners, all gathered to witness their king declare his dominion over the heavens.

Amidst the grandeur, a small boy entered the hall. Barefoot, with a simple staff in hand, the child's unassuming presence seemed almost laughable against the pomp of the occasion. Yet, when he spoke, the hall fell silent.

'O King Bali,' he said. 'I have heard of your generosity. I seek a gift, not gold, jewels, or cattle. I ask for three paces of land, measured by my stride.'

A murmur rippled through the court. Some chuckled; others whispered, 'Why waste a boon on this boy?' But Bali, ever the giver, smiled. 'Three paces of land? It shall be yours.'

Shukracharya, his guru, stepped forward, his voice urgent. 'Bali, stop! This is no ordinary boy. He is Vamana, the divine. His request is a test, and your promise will undo you.'

Bali paused, torn between the words of his guru and the promise he had made. Finally, he said, 'If it is a test, then so be it. What is a ruler without his word?'

The First Step: The Illusion of Ownership

Vamana smiled and placed his first step. The Earth trembled. Rivers reversed their flow, mountains bowed, and Bali watched in awe as the boy's foot spanned the entire Earth. For the first time, Bali felt the weight of his own hubris. He had ruled the land, but did he own it? The soil beneath his feet had nurtured life long before his reign. His people had tilled its fields, and nature had shaped its beauty. What had he truly given to it, beyond his name?

The Second Step: The Limits of Ambition

Vamana lifted his foot again, this time stretching into the heavens. The skies darkened as the boy's stride covered the stars, the sun, and the moon. Bali's vision blurred, his dreams of ruling the cosmos vanishing like mist. In that moment, Bali understood the fragility of ambition. The heavens had never been his to claim. They were a shared inheritance—a canvas painted by countless forces beyond his comprehension.

The Third Step: The Surrender of Self

Vamana turned to Bali. 'O king, where shall I place my third step?'

Bali knelt, his heart heavy, 'On my head, Lord. I surrender myself, for I see now that my pride blinded me to the truth.'

As Vamana's foot touched Bali's crown, Bali felt a strange liberation. In giving up all he had sought to hold, he discovered the truth: the universe is not a prize to be won but a balance to be cherished.

Layers of Meaning

- **Human free will and its duality:** Bali's journey reflects the eternal struggle of free will. It is our greatest gift, allowing us to dream and create, yet it can lead us astray when untethered from humility and Dharma.
- **The illusion of possession:** The story asks us to question what we truly own. The Earth, the air, the water—these are not possessions, but sacred trusts. Bali's surrender reminds us that stewardship is more meaningful than dominion.
- **The balance of power:** Vamana's steps symbolise the cosmic scales. When humans tilt the balance—be it through over-extraction, inequality, or unchecked ambition—the universe intervenes, sometimes gently, sometimes forcefully.

Whisper for the Future

In today's world, Bali's tale is a mirror to our times. We drill deeper, build higher, and claim more, forgetting that the Earth does not belong to us alone. Our free will has led to technological marvels but also ecological collapse.

The story calls us to pause and ask:

What is our third step?

Perhaps it is time to kneel, not in defeat but in humility, and let the Earth reclaim its balance. For in surrendering our pride, we may rediscover our place in the grand design of creation.

Silence of the Bow: Lessons in Ahimsā

It was the eve of war. The battlefield of Kurukshetra stretched like a restless sea under the twilight, shimmering with the anticipation of chaos. Arjuna, the warrior prince, stood at the centre of the storm, his chariot drawn by celestial horses, his bow, Gandiva in hand. Yet, he was unarmed—not in weapons, but in will.

'I cannot do this, Krishna,' Arjuna whispered, his voice trembling. 'How can I raise my hand against my own kin? What victory is worth this destruction? What Dharma guides a hand that wields death?'

Krishna, his charioteer and the eternal teacher, listened patiently. His smile held the wisdom of ages. 'Arjuna,' he said, 'the battle before you is not just of arrows and swords. It is a battle within—a conflict between duty and compassion, between free will and divine order.'

Moment of Choice

Krishna's words hung in the air as Arjuna gazed at the horizon. The Kauravas stood as shadows, brothers turned enemies by greed and pride. The Pandavas, his own family, stood behind him, their trust heavy on his shoulders.

'Tell me, Krishna,' Arjuna pleaded, 'where does *Ahimsā* find its place in this war? Is it in standing silent, or in raising my bow to fight? Am I to choose between slaughter and cowardice?'

Krishna's eyes twinkled, as if the universe itself danced within them. '*Ahimsā*, Arjuna, is not the absence of action. It is the absence of malice. It is the intent that defines violence—not the act itself. This battle is not about conquest but about restoring balance. Yet, the choice remains yours. Free will is both a gift and a test.'

Layers of Conflict

Arjuna's fingers tightened around Gandiva as he struggled with Krishna's words. He thought of Bhishma, his grandsire, who had once cradled him as a child but now stood against him. He thought of Karna, his half-brother, bound by loyalty to the Kauravas. Each arrow he shot would pierce not just bodies, but bonds.

Yet, Arjuna also saw the pain of Draupadi, his queen, humiliated before a silent court. He saw the farmers, the weavers, and the labourers crushed under Duryodhana's tyranny. Ahimsā, he realised, was not a passive ideal. It was a responsibility—a force that demanded courage and discernment.

Cosmic Vision

Krishna, seeing Arjuna's turmoil, revealed his universal form—the **Vishwaroopa**. In that moment, Arjuna saw time unravel: the birth of stars, the destruction of worlds, the dance of creation and dissolution. He saw the battle itself as a single thread in an infinite tapestry.

'I am Time,' Krishna said, 'the destroyer of worlds. Yet, I am also the preserver. Understand, Arjuna, that Ahimsā is not mere restraint. It is alignment with the Dharma of the universe. Violence born of hatred binds you, but action born of duty liberates you. The choice is yours.'

Loss and Renewal

Arjuna's vision cleared. He raised Gandiva, not with pride but with purpose. Each arrow he released carried not anger but a prayer—for justice, for balance, for the countless lives yet to come.

The war raged, and with it came loss, but also renewal. The soil of Kurukshetra, soaked with blood, would later bloom with the seeds of peace. Arjuna fought not to destroy but to preserve the Dharma that held the universe together.

Layers of Meaning

- **Ahimsā as intent, not inaction:** The story reveals that Ahimsā is not about avoiding conflict but about ensuring that one's actions are rooted in love and justice, not hatred. Even in war, Arjuna's aim was to restore harmony, not to destroy for personal gain.
- **The role of free will:** Arjuna's choice to fight was not compelled by Krishna but guided by his own understanding of Dharma. Free will allows us to navigate complex moral dilemmas, balancing personal values with the collective good.
- **The balance of creation and destruction:** The Vishwaroopa reminds us that destruction is a part of life's cycle. Ahimsā does not mean denying this reality but participating in it with mindfulness and compassion.

Modern Reflection

In today's world, Ahimsā challenges us to act responsibly in the face of climate change, inequality, and injustice. It asks us to confront hard truths and take difficult actions—not out of anger but out of love for future generations.

Whisper for the Future

Arjuna's story is not confined to the battlefield. Each of us stands on our own Kurukshetra, facing choices that ripple through time. Do we act out of fear and hatred, or do we embrace Ahimsā, using our free will to restore balance? The Earth watches, as Krishna once did, waiting for us to raise our bow—not in violence but in defence of harmony.

Pancha Mahabhūtas: The Five Elements

Long ago, in Rishikesh on the banks of the River Ganga, people lived in harmony with nature, each individual attuned to the rhythm of the Earth. They understood the sacred balance of life through the principle of the *Pancha Mahabhūtas*, the five elements—Earth, Water, Fire, Air, and Space—that formed the foundation of all existence. The village was ruled by King Raghavendra, who, like his ancestors, had deep respect for nature and its teachings. One day, as he meditated by the river, a voice called out to him from the sacred waters. It was a Yaksha, a mystical spirit of the forest, who had witnessed the waning balance of the elements in the world.

Challenge of Balance

The Yaksha's voice resonated with the winds and the Earth itself: 'King Raghavendra, the harmony of the five elements is in peril. The Earth is being stripped of its forests; the rivers are drying up, poisoned by the hands of humankind; the Fire that once illuminated life now consumes it; the Air grows heavy with dust and fumes; and the Space between all beings is shrinking. The balance of the *Pancha Mahabhūtas* has been disrupted, and soon, this world will teeter into chaos. You, as a guardian of this land, must choose your path.' Raghavendra, filled with both fear and resolve, replied, 'I understand the sacred dance of these elements, but what must I do? The world is changing, and our ways seem too small to alter the course of this destruction.'

Lesson from the Past

The Yaksha smiled, its form flickering between light and shadow. 'You are right, King. The balance of the *Pancha Mahabhūtas* has always been delicate. To understand your role, let me share the tale of an ancient king, Sage Manu, whose kingdom thrived because he understood the power of the elements.'

The Yaksha began: 'Long ago, Manu ruled over a prosperous land where nature was revered. One day, a great flood threatened to consume everything. Manu sought guidance from the wise sage Vishvakarma, who revealed to him a vision of the future: the Earth would flood, but Manu's wisdom would guide his people to salvation.'

'Manu,' Vishvakarma said, 'the balance of the *Pancha Mahabhūtas* must be maintained. You must plant seeds for the Earth's fertility, preserve Water's purity, control Fire's rage, nurture the Air's flow, and honour Space as a sacred void between all beings. If these elements are preserved, the flood will recede and life will continue. If not, destruction will follow.'

Manu heeded the sage's words. He planted trees to preserve the Earth, created channels to purify and conserve Water, built hearths to control the Fire, kept the Air fresh with herbal gardens, and ensured that Space was free for the movement of life, understanding that cluttered lives create discord. The flood came, but Manu's kingdom survived because he had learned to honour and balance the five elements. The Earth was fertile, the Water was clear, Fire was controlled, the Air was pure, and the Space between all things was honoured. He had understood that humanity's free will could either restore or disrupt the balance of nature.

Eternal Choice of Free Will

As the Yaksha's tale ended, it returned to its cryptic stillness. 'Raghavendra, you stand at a crossroads. The lessons of Manu hold the key. The free will of humankind has brought this imbalance, but the same free will can restore it.'

Raghavendra bowed his head in contemplation, for he knew that the path forward was not merely about action but about awareness. He understood that in every choice, the elements were at play. To heal the Earth, he would have to reconnect people with the ancient wisdom that had long been forgotten.

The Awakening

The King called for an assembly, not just of his courtiers, but of the common folk—farmers, artisans, traders, and even the children. He spoke to them of the *Pancha Mahabhūtas* and their sacred balance.

'Just as the Earth gives life, so must we nurture it. Just as Water is the source of all, so must we purify and conserve it. Fire is the force of transformation, but it must be controlled, for it can also destroy. The Air is the breath of life, and we must keep it untainted for future generations. Space, the empty canvas of existence, must be honoured for it creates room for harmony.'

Raghavendra then set forth a plan to restore balance:

- **Earth (*Prithvi*)**: He initiated massive tree-planting campaigns to restore the forests, replenishing the soil's vitality. He promoted sustainable farming practices that honoured the Earth's ability to regenerate.
- **Water (*Jal*)**: The rivers were cleaned, and an innovative systems for water conservation were introduced. The villagers built rainwater harvesting structures to ensure water was available for generations to come.
- **Fire (*Agni*)**: Controlled fires were used for warmth and cooking, while a deep understanding of fire's role in transformation led to better agricultural practices and a reduced reliance on destructive fires.
- **Air (*Vāyu*)**: The village began planting herbal gardens to purify the air, reducing pollution. The people also adopted cleaner energy sources, reducing their carbon footprint.
- **Space (*Ākāśa*)**: The king encouraged the creation of open spaces for the people, promoting mental clarity and peace. They built communal spaces where individuals could connect with nature and with each other.

A Call to the Future

King Raghavendra's village flourished as it had once, guided by the wisdom of the *Pancha Mahabhūtas*. However, the king did not rest on his laurels. He knew that the balance of the five elements was fragile, and it was a lesson for the future—a lesson that must be passed down through generations.

'Look around you, my people,' he said. 'The Earth is not just a resource, but a living entity. Water is not just for drinking, but the essence of life itself. Fire is not to be feared, but respected. Air is the life-giver, and Space is the sacred pause in which life can grow. Remember that we are not separate from the world around us. We are its stewards, not its conquerors. We must use our free will wisely, for the actions of today will shape the world of tomorrow.'

And so, as the years passed, the wisdom of the five elements became the foundation of not just the village, but an entire kingdom that spread the message of balance, sustainability, and respect for the natural world. The lesson of King Raghavendra echoed down the centuries, a timeless reminder that humanity's relationship with the Earth is a delicate dance, and it is our responsibility to maintain its harmony.

Layers of Meaning

- **Free will and responsibility**: Just as Raghavendra chose to follow the wisdom of Manu, each person has the power to act, but with this power comes the responsibility to maintain balance. Free will must be exercised with an understanding of the consequences on the *Pancha Mahabhūtas*.
- **The balance of the five elements**: The story reveals that no element can thrive in isolation. Earth, Water, Fire, Air, and Space must exist in harmony. Similarly, the modern world must address the ecological imbalances that arises from unchecked exploitation of resources.
- **Sustainability and harmony**: The lesson of Manu and Raghavendra speaks of sustainability as a conscious, collective effort. Sustainable urban planning, renewable energy, and conservation efforts are the modern-day applications of the ancient wisdom of the *Pancha Mahabhūtas*.

- **A future of balance**: The choices we make today—regarding climate change, resource depletion, and pollution—will determine the world of tomorrow. We must act with the wisdom to restore and maintain the balance of the five elements, for the Earth is not merely a resource but a partner in life.

Whisper for the Future

As the rivers flow and the winds shift, may we remember the lessons of the past. The Earth is our mother, Water our lifeblood, Fire our transformation, Air our breath, and Space the canvas of life. To restore balance, we must live with awareness, compassion, and respect for the elements that sustain us. Our free will holds the power to heal or harm. What will we choose? The Earth is waiting for our answer.

Annam Brahma: The Sacredness of Food

Long ago, in Kurukshetra, where the battle of the Mahabharata was to unfold, a small village existed nestled between rolling hills and fields of abundant grain. The villagers, humble in their ways, had always lived by the principle that *Annam Brahma*—food is divine. They considered food to be the very essence of life, a gift from the Earth and the divine forces that governed it. The village had been thriving for centuries, not because of its wealth or military power, but because it had understood the deep spiritual connection between food, nature, and the divine.

The villagers' rituals reflected their reverence for food. They gave thanks to the Earth before every meal, shared their harvest with the poor, and never allowed even the smallest morsel to be wasted. But a storm was coming, one that no one could foresee, a storm that was not made of rain or wind. It was the storm of human disregard for the sacredness of food, a storm that would uproot their way of life.

Call of the Divine

One evening, as the sun dipped below the horizon, Rishi Agastya, a sage known for his wisdom and connection to the divine, appeared in the village. The villagers gathered at his feet, seeking his guidance, for the skies had darkened and strange signs had begun to appear. Agastya, with a calm and piercing gaze, addressed the villagers: 'The balance you cherish—your reverence for food as the divine gift—is under threat. The principles you live by are being forgotten, for in far-off lands, people consume without awareness, waste without thought, and abandon the sacredness of food in pursuit of their own desires.' The villagers, puzzled and frightened, asked, 'But how can this be? We honour food every day. We share, we nurture the land, and we waste nothing. How can we protect what we hold sacred?' Agastya's eyes softened. 'Your ways are pure, but you must remember, the forces of human free will are strong. People are beginning to forget the sanctity of food. When food is wasted, it is not just a loss of sustenance; it is a loss of grace. The divine is insulted. And when the divine is insulted, nature itself begins to wither.'

Story of King Harishchandra

To explain his words, Agastya recited a story from the ancient past. 'Long before this age,' the sage began, 'there was a king named Harishchandra, renowned for his integrity and righteousness. One day, Harishchandra decided to test his virtue by giving away everything he had, from his riches to his lands, to the poor and the needy. Yet, despite his generosity, he could not escape the inevitable hardships of life. He gave away his food, his treasures, and even his kingdom, living in a state of complete destitution.

'In his darkest hour, while sitting beneath a tree with nothing but the remnants of a meal he had gathered from the forest, a celestial being appeared. The Being asked, "Why are you so upset, oh King? You have given everything away, yet you still carry the weight of sorrow in your heart."

'Harishchandra replied, "I have given away everything, but I still find no peace. I have even given away my last morsel of food to a hungry traveller, yet my spirit remains restless."

'The celestial being smiled and said, "It is not in giving away your food that you will find peace, but in understanding that food is not simply sustenance—it is the life-force of all beings. To waste it is to dishonour the divine. To honour it is to honour the very nature of existence."

'At that moment, Harishchandra understood. It wasn't the loss of wealth or possessions that troubled him. It was the failure to recognise that food, as the divine gift of nature, must be treated with reverence. He returned to his people, restored, and rebuilt his kingdom with the understanding that food was not to be consumed thoughtlessly but must be respected as the divine energy of life itself.'

Power of Human Free Will

As Agastya finished his story, the villagers were moved, understanding the gravity of the situation. The sage continued, 'Harishchandra's story reminds us of the power of human free will. You see, it is the free will of human beings that determines whether food is consumed in reverence or discarded carelessly. In your world today, food is no longer seen as sacred. The rising forces of greed, excess, and waste threaten the balance of nature.'

The villagers nodded sombrely, realising the truth in the sage's words. Agastya looked at them and said, 'The divine is not merely in the act of offering food to the gods. It resides in the very act of mindful consumption. *Annam Brahma* must be understood at a deeper level. When food is consumed with gratitude, shared with others, and not wasted, it reflects the divine flow of energy from Earth to humanity, and back to Earth again.'

Balancing the Future

Agastya then laid out a vision for the future. He spoke of the human free will as the bridge between respect and disrespect for the divine in food. He emphasised that the imbalance caused by the over-consumption and waste of food would lead to further ecological damage and suffering for generations to come.

'The Earth,' he said, 'is a living being, and food is its life force. In consuming, we must remember that we do not just consume to satiate ourselves, but to nourish the Earth as well. When food is wasted, it is the Earth that suffers. When food is respected, it is the divine that is honoured.'

He then instructed the villagers: 'Plant your crops with care, consume with mindfulness, and ensure no food is wasted. Let your lives reflect the understanding that food is sacred and to waste it is an insult to the divine.'

Legacy of the Sacred Grain

The villagers took the sage's teachings to heart. They transformed their land into a model of sustainable agriculture. They never planted more than the land could yield, and always shared their bounty. Their meals were prepared with reverence; every bite consumed with gratitude. Food was never thrown away; leftovers were given to the poor, and nothing was left to spoil.

The story of the village's transformation spread far and wide, a beacon for future generations. As cities grew and populations increased, people began to heed the lessons of the village. They planted organically, reduced food waste, and created systems to share excess food with those in need.

Mindful eating became a practice, and the sacredness of food was taught in schools and households alike. The lesson was clear: *Annam Brahma* is not merely a belief; it is an action, a way of life. We are the stewards of the Earth, and the food we consume is not our possession alone, but the gift of nature and the divine.

Future Lessons

In the future, the lessons from Agastya's teachings would continue to resonate:

- **Respecting food as divine**: Understanding that food is not merely fuel but the essence of life; it is something that we must consume with reverence, respect, and mindfulness, ensuring that no morsel goes to waste.
- **Human free will in action**: The choices we make daily, in how we grow, consume, and discard food either honours or dishonours the sacred gift of sustenance. Our free will carries the power to restore balance or perpetuate imbalance.

- **Balance in consumption**: Just as Harishchandra learned, the act of mindful consumption is an offering to the divine. Excessive consumption, driven by greed or wastefulness, disrupts the balance of nature and harms the divine connection between Earth and humanity.
- **Sustainability through stewardship**: The Earth is not a resource to be exploited, but a living being to be cared for. Sustainable agricultural practices, food waste reduction, and sharing the bounty are ways in which we honour the divine principle of *Annam Brahma*.

Whisper for the Future

As the seasons change, may we remember that food is not just sustenance; it is sacred. Every grain that grows from the soil, every meal prepared with love, and every act of sharing is a reflection of the divine. In this sacred act, we find not only nourishment for the body, but for the soul. Let us live with the understanding that to waste food is to waste the divine, and to honour it is to honour the Earth itself.

Karma Yoga: Power of Selfless Action

In ancient Dharanapura, a small village known for its verdant fields and peaceful rhythm of life, lived a humble farmer named Yashwanth. He was a man of few words, but his actions spoke volumes. While others around him fretted over the demands of life—growing crops, tending to cattle, managing their homes—Yashwanth went about his duties with unwavering focus, never distracted by the results. He laboured day and night, cultivating his fields with love, caring for his animals with the same devotion, and offering his produce to the village without hesitation.

The villagers often marvelled at Yashwanth's ability to work tirelessly, yet remain serene and content despite the unpredictability of the harvest. They would ask him, 'How do you manage to maintain such peace? Do you not worry about the future? Do you not yearn for the fruits of your labour?'

Yashwanth would smile softly and reply, 'I do not worry about the outcome because my only duty is to act with integrity. The rest is beyond my control.'

A Moment of Crisis

One day, a great calamity struck the village. The skies darkened, and the rains did not come. Drought set in, causing the crops to wither and the fields to dry. The villagers were despondent. They feared for their survival, as they relied heavily on the harvest for sustenance. Yashwanth, however, continued his work undeterred. He irrigated his fields with the little water he had, carefully tended to the last remaining crops, and helped his neighbours where he could.

When asked why he did not worry about the disaster, he responded simply, 'What is the use of worrying about the storm when I can only do what is in my power? I shall continue working, as I always have.'

But the villagers were puzzled. 'Yashwanth, don't you see? The crops are failing; the rains have not come. Your work will amount to nothing. Why continue if the future is uncertain?'

Yashwanth looked at them with quiet wisdom, and then, after a pause, he told them a story.

King Janaka's Story

'There was once a great king,' Yashwanth began, 'whose name was Janaka, a ruler of immense wisdom and righteousness.

One day, a wandering sage arrived at the king's court. This sage, whose name was Ashtavakra, was known for his knowledge of the self and the universe.'

'Ashtavakra asked the king, "Oh King, you rule over many lands, yet are you attached to the fruits of your rule? Do you govern with the expectation of personal gain, or do you see your actions as a duty?"

'King Janaka, who had long ago transcended personal desires, smiled and replied, "I rule because it is my duty. I do not seek fame, nor wealth, nor personal glory. I act because it is right, and the results are not for me to control. I sow seeds in the field of Dharma, not in the soil of selfishness."

'Ashtavakra, impressed with the king's wisdom, said, "You understand, oh King. The path of action without attachment is the path of the truly free".'

Wisdom of Karma Yoga

Yashwanth paused for a moment, as the villagers listened intently to his tale. 'The lesson of King Janaka is clear,' he continued. 'He did not rule for personal gain. He ruled because it was his Dharma, his duty to his people. And like him, I labour not for the fruit of my work, but because it is my responsibility to nurture life, to care for the land, and to do so with all my heart. Whether the rains come or not, I continue my work.'

'The principle of Karma Yoga,' Yashwanth explained, 'teaches us that it is the action that matters, not the result. It is the work itself that is sacred, for it is our duty to care for the world we inhabit. The outcome is not in our hands, and the fruits of our labour are not ours to claim.'

Crisis and Turning Point

As days passed, the crisis grew more dire. The villagers, unable to cope with the drought, became increasingly despondent. Yet, Yashwanth continued with his work, unaffected by the apparent failure of his efforts. One day, a traveller arrived in the village, weary from his journey. He asked the villagers, 'What is happening here? Why do you look so troubled?'

'The rains have not come,' they explained. 'Our fields are dying, and our people are hungry. It seems there is no hope.'

The traveller, a wandering sage, turned to Yashwanth and asked, 'You, who are working with such dedication, even though the rains have not come—why do you continue?'

Yashwanth smiled and responded, 'I continue because I must. I have no power over the weather, but I do have control over my actions. I do not work for a reward; I work because it is my duty. This is the essence of Karma Yoga—action without attachment.'

The traveller nodded thoughtfully and then revealed that he had the power to summon the rains. 'But,' he said, 'I will not do so unless you, Yashwanth, can show me that you act with true selflessness, without concern for the outcome.'

Yashwanth, without hesitation, replied, 'I do not seek a reward. My labour is my offering to the Earth. If the rains come, it will be a blessing. If they do not, I will continue my work, regardless.'

The traveller, moved by Yashwanth's steadfastness, smiled and raised his arms to the sky. Within moments, the clouds parted, and the rains began to fall.

Lesson for the Future

As the village rejoiced in the arrival of the rains, Yashwanth turned to the villagers and said, 'The rains have come not because I asked for them, but because I did what was right, without expecting anything in return. This is the power of selfless action, of Karma Yoga.'

He explained further, 'In our world today, we are faced with challenges that seem insurmountable—environmental degradation, climate change, and the depletion of resources. Many may wonder, "Why act when the world is so broken? What difference can my small actions make?" But the answer lies in the essence of Karma Yoga. It is not the outcome that defines our actions, but the integrity of the work itself.'

'Each step we take towards sustainability, no matter how small, is an act of duty. It is our responsibility to the Earth, to future generations, and to ourselves. And though we may not see immediate results, we must act with the same unwavering commitment as I do in tending to my fields—because it is the right thing to do.'

Ripple of Action

The villagers, inspired by Yashwanth's words, began to change their approach to the world around them. They no longer focused on immediate gains, but instead, embraced the principle of Karma Yoga in all their actions.

They cared for the land, conserved resources, and shared their harvests, not for recognition or reward, but because it was their sacred duty. And as they did so, the village began to thrive—not only because the rains came, but because their hearts had been awakened to the power of selfless action.

Future Vision

In the future, this lesson would echo in the hearts of all

those who sought a more sustainable world. The principle of Karma Yoga would guide individuals and organisations alike, to take responsibility for the environment, to act without attachment to results, and to focus on ethical behaviour and collective well-being.

Just as Yashwanth's actions were not swayed by the fear of failure, so too must humanity continue to act in the face of environmental challenges, knowing that each small step, each mindful action, contributes to the greater good. By embodying the essence of Karma Yoga, we can restore balance to the Earth and ensure a sustainable future for generations to come.

Whisper for the Future

In the quiet moments of our lives, when we feel overwhelmed by the enormity of the world's problems, let us remember Yashwanth's steadfastness. Let us act with selflessness, knowing that the work we do, no matter how small, has its place in the greater cosmic order. For it is in the purity of our actions, untainted by desire for reward, that we can find true balance—and in turn, help restore balance to the Earth.

***Prakṛti Rakṣā:* The Sacred Forest of Satyapada**

Long ago, in a kingdom called Satyapada, where the rivers flowed with crystal-clear waters and the air was fragrant with the blooms of endless forests, there stood a sacred grove deep within the heart of the land. This grove, known as the Kailasa Vana, was believed to be protected by the goddess Vishala, a divine spirit of the forest. The grove was home to trees that had stood for centuries, their roots deeply entwined with the Earth itself, and their leaves whispered ancient secrets to those who listened carefully.

The people of Satyapada had lived in harmony with the land for generations, guided by the belief that nature was divine and sacred. They treated the trees, rivers, and animals as their kin, offering their prayers, preserving their habitats, and protecting them from harm. It was a balanced relationship—humans took only what was needed and gave back with reverence and respect. The people believed that as long as the forest thrived, the kingdom would prosper.

A Time of Reckoning

However, as with many stories of old, greed and ambition began to seep into the hearts of some. Chandran, a young and ambitious ruler who inherited the throne of Satyapada, sought to expand his kingdom's borders. He was wise in many ways but lacked the reverence for nature that his ancestors had cherished. He believed that the forest was an obstacle to progress, and that the resources it contained could be used to build grand palaces, markets, and roads. The idea of exploiting the sacred grove for timber and precious resources began to grow in his mind.

Chandran convened his council of advisors and declared, 'The Kailasa Vana stands in the way of our prosperity. Let us clear the land and use its timber to build an empire that will last for millennia. This kingdom must grow, and for it to grow, we must use the resources that lie within this sacred grove.' The elder sage of the kingdom, Raghavan, stood up in protest. He had witnessed the harmony between humans and nature for decades and knew the consequences of upsetting that balance. He spoke with a gentle but firm voice. 'Your Majesty, the forest is not just land. It is sacred. It is the heart of Satyapada. To disturb it is to disturb the divine harmony that sustains us.

'Goddess Vishala herself watches over these woods. If we harm this place, we will upset the delicate equilibrium that has kept our land fertile and our people strong.' But Chandran, swept by his ambition and the pressure of his advisors, dismissed Raghavan's concerns. 'Times have changed. The world is evolving, and we must evolve with it. We must take what is ours to ensure our future,' he retorted. And so, the king ordered the forest cleared.

Repercussions of Disruption

In the days that followed, the mighty trees of the Kailasa Vana were felled, one by one, their ancient trunks turned into timber for new buildings. The sacred river that flowed through the grove was redirected to irrigate the expanding fields. The animals fled in fear, and the once-vibrant land began to wither.

The harmony that had once existed between humans and nature was broken, and with it, the kingdom began to falter. Within weeks, the soil lost its fertility. Crops withered, the river became polluted, and the air grew thick with dust. Famine struck, and the people of Satyapada began to suffer. The kingdom, once prosperous and peaceful, was now on the brink of collapse.

Raghavan, who had warned the king, watched in sorrow as the consequences of the disruption became evident. He went to the royal palace, determined to speak to the king once more.

A Humbling Visit

When Raghavan arrived before King Chandran, the once-proud ruler now looked weary. His eyes were sunken, and the burden of his decisions weighed heavily on him. 'Raghavan,' the king said, 'I have seen the destruction that has followed my orders. Our kingdom is crumbling. What must I do to restore balance? I see now the folly of my actions, but how can I undo the damage I have caused?'

Raghavan, filled with compassion for the troubled king, replied softly, 'It is not enough to simply mourn the past, Your Majesty. Nature's balance cannot be restored through regret alone. You must act with humility and responsibility. Just as you once used your free will to disturb the balance, now you must use that same free will to heal it.' The sage continued, 'Goddess Vishala is merciful, but she requires that we offer her respect, humility, and dedication. Only then can we begin to heal the wounds we have inflicted upon the land.'

The Path of Restoration

Chandran, realising the depth of his mistake, summoned his people to a great assembly. 'I have wronged the land and the divine forces that protect it,' he said. 'Now, I seek your help. Let us return to the old ways, to the ways of balance and reverence for nature. Together, we shall restore what was lost.' Under Raghavan's guidance, the people of Satyapada began the slow and arduous task of healing the land.

They replanted the trees that had been felled, sowed seeds to replenish the soil, and restored the flow of the river to its natural course. They enacted laws to protect the forests, to conserve water, and to treat animals with respect. They offered prayers to Vishala, asking for forgiveness and promising to live in harmony with the land once more.

Lessons for the Future

Over time, the kingdom began to heal. The soil regained its fertility, the river flowed clear again, and the forests slowly returned. The people of Satyapada understood that their free will—though a powerful force—was not to be wielded recklessly. They learned that nature, in all its beauty and majesty, was a gift from the divine, and that to live in harmony with it was not a choice, but a sacred responsibility.

And so, the kingdom flourished again, but this time with a deep understanding of the balance that must be maintained between human progress and the protection of nature.

The Ripple Effect: *Prakṛti Rakṣā* in the Modern World

The story of Satyapada offers us a timeless lesson. Just as King Chandran used his free will to disrupt the harmony of the land, modern society often exploits the natural world in the name of progress and growth. But like the people of Satyapada, we too must learn to live in balance with nature, to protect the sacred forests, rivers, and mountains that sustain us.

Our future depends on the choices we make today. If we continue to take from the Earth without giving back, we risk the same fate that befell Satyapada—destruction and imbalance. But if we act with responsibility, with reverence, and with a commitment to *Prakṛti Rakṣā,* we can create a future where both humanity and the natural world thrive together in harmony.

The choice is ours.

Aranya: The Whisper of the Forest

In ancient times, there was a village where people lived in deep harmony with the land and the forest. The village followed the teachings of Aranya, which held that nature was not merely a resource but a sacred partner in life. The people believed that to live was to be in a constant dialogue with nature, to listen to its whispers, and to act with reverence. Every tree, every river, every animal was seen as part of a grand design, and the forest was not just a physical space—it was a reflection of divine harmony.

The villagers were mindful of their consumption. They harvested only what they needed, never more. The women would gather herbs, but only those that had ripened naturally; the men hunted, but only what was in abundance. Nothing was taken without care. They lived simply, following the principle that to take from nature was a privilege, not a right, and that balance must be maintained. At the heart of this village lived an elderly sage, Vaidya, who had devoted his life to studying the deeper meaning of the forest. His life was one of *tapasyā*, austerity, self-discipline, and profound restraint. Vaidya understood that the forest was not just a source of sustenance, but a sacred entity—an embodiment of *Prakriti*, the natural world, where every element worked together in perfect balance.

The Dilemma of Young Sumantra

One day, a young man named Sumantra returned to the village after many years of travelling. He had seen the ways of distant lands, cities that pulsed with energy and progress, where machines harvested crops and factories produced goods in abundance. These places were full of people with desires that seemed insatiable, always striving for more.

When Sumantra met Vaidya, he was filled with excitement. 'Vaidya,' he exclaimed, 'I have seen great cities where people live in luxury, surrounded by abundance. They have more than they could ever need—machines that work for them, goods that are endlessly produced. Why must we live so simply, so quietly, in this forest? Why should we deny ourselves the comforts of progress?' Vaidya listened carefully, then, without immediate response, he invited Sumantra for a walk through the village's surrounding forest.

A Walk Through the Forest

As they walked, Vaidya spoke, 'Sumantra, look around you. Do you see how the trees stand tall and strong? The rivers flow gently, the soil is rich, and the animals move freely through this place. This forest is not just a collection of trees and plants; it is a living entity, a community where each being plays its part in maintaining harmony.' Sumantra looked around, noticing the peaceful co-existence of all things. 'But surely, the world is calling us to progress. The cities are growing, and with them, the opportunities. Why should we not seek to evolve as they have?'

Vaidya smiled, his voice calm. 'The question is not whether progress is wrong, but whether it comes at the cost of balance. The cities you speak of are built on unrestrained desire. The land is exploited, the resources drained, and in the end, the imbalance that follows leads to depletion. The trees that once stood tall are now gone, the rivers poisoned, and the soil infertile.'

Sumantra frowned. 'But isn't that the price of advancement? Should we not aspire to create more, to build faster, to have everything?'

Vaidya's gaze softened. 'Advancement, yes, but with wisdom. The forest does not take more than it needs, yet it gives endlessly. The animals eat only to satisfy their hunger, and the trees shed leaves to return nutrients to the Earth.

Each part of the forest is in balance, and that is why it thrives. This is *Tapasyā*, living with self-restraint, knowing when to take and when to leave. True growth does not come from excess but from a balanced approach to life.'

Lessons for the Future: Living with Balance

Sumantra spent the next few days contemplating Vaidya's words. Slowly, he began to understand that the wisdom of the forest, the principles of Aranya, were not about rejecting progress but about ensuring that it did not come at the expense of balance. The future of the world could not rest on unchecked consumption; it had to be built on wisdom, restraint, and an understanding that humans were not separate from the Earth but a part of it.

As Sumantra meditated under a canopy of trees, he realised that self-discipline was not just an individual choice but a collective responsibility. The people of the forest had lived for generations in harmony with nature because they knew that the health of the land depended on their actions. If humanity were to continue evolving, it must do so with a sense of balance—a balance that honoured the Earth. In today's world, the story of Aranya teaches us a vital lesson: we must remember that our free will, our desire for progress, comes with a profound responsibility to protect and preserve the balance of nature.

As we strive for growth, we must ensure that we do not deplete the resources that sustain life. Just as the forest thrives through careful restraint, so too must we choose wisely what we consume and how we act, recognising that our actions impact the delicate web of life around us. The future rests in the choices we make today, and these choices should reflect the wisdom of the forest—where balance, harmony, and respect for all beings are the foundations upon which true progress is built.

Tapasyā: The Concept of Minimalism

Sage Vasishtha's abode, on a remote hill in the dense forest, was a simple hut made from the surrounding elements—wood, leaves, and clay—the materials he had gathered in harmony with nature's rhythms. He neither sought nor needed the comforts of the world, for his life was an embodiment of *Tapasyā*—a disciplined, austere way of living that led to inner peace and connection with the divine. One morning, a young prince, Raghav, weary of his lavish life, sought Vasishtha's wisdom. He had inherited vast wealth, magnificent palaces, and an army of servants. Yet, none of these material possessions could fill the growing void in his heart. He was disturbed by the unrelenting cycle of desires that never seemed to be quenched. The more he acquired, the more he craved.

In search of answers, Raghav ventured into the forest to meet the sage. When he arrived at Vasishtha's humble hut, the sage was deep in meditation. The prince waited patiently until the sage opened his eyes. 'I have come with a question, venerable sage,' Raghav said, his voice heavy with the burden of unfulfilled desires. 'I have everything a man could dream of—wealth, power, and prestige—but I feel empty. Why is this so, and what can I do to find true contentment?' Vasishtha looked at the young prince, and after a long pause, he spoke: 'Tell me, young prince, what is it that you truly desire?'

Raghav hesitated. He was not used to such direct

questions. He had been taught to want more—to conquer, to expand, to never be satisfied. 'I desire peace, freedom from this endless cycle of craving. I wish to know the truth, to understand why despite everything I have, I am still not content.' The sage nodded and said, 'Then let me tell you a story from our ancient scriptures, from the time when even the gods had to confront their desires.'

King Rantideva's Story

There was a king named Rantideva who ruled over a prosperous kingdom. King Rantideva had everything—a flourishing kingdom, loyal subjects, and immeasurable wealth. Yet, his heart was troubled. Despite having all the comforts and luxuries, he realised that true peace could not be bought with riches. His desire for peace, harmony, and ultimate fulfilment grew stronger.

One day, while walking in his gardens, Rantideva encountered a beggar. The beggar, tired and weak, asked for alms. King Rantideva, moved by compassion, offered the beggar everything he had on him—food, water, and gold. The beggar, in turn, expressed gratitude and shared his humble wisdom: 'True fulfilment comes not from having more, but from giving more.'

As King Rantideva pondered these words, his kingdom was struck by a severe drought. Crops failed, rivers ran dry, and famine spread across the land. Yet, Rantideva remained undeterred. He continued to practice generosity, giving away every resource his kingdom had. He offered food and water to the needy, even as his own family went without.

With each act of selflessness, his heart became lighter, and the burden of desires began to lift. In time, the gods noticed King Rantideva's unwavering commitment to selfless service and austerity. They appeared before him and granted him divine wisdom and peace. King Rantideva had learned the lesson of *Tapasyā*—that true fulfilment arises not from indulgence, but from self-restraint, simplicity, and disciplined living.

Vasishtha's Lesson

As the story concluded, Sage Vasishtha turned to Prince Raghav. He spoke with quiet intensity: 'King Rantideva's life was the embodiment of *Tapasyā*. He did not allow his desires to control him, nor did he let the world dictate his happiness. He lived simply, yet with immense generosity, for he understood that austerity does not mean deprivation, but rather, a conscious choice to live with less, in balance with nature and others.'

The sage paused, allowing Raghav to reflect on the story. 'The world today is filled with desires, much like the kingdom of Rantideva. People consume endlessly, believing that more will bring happiness. But just as the drought in Rantideva's kingdom revealed the true value of generosity and self-discipline, so too must we see the consequences of our modern desires. The world's resources are finite, and our consumption has reached unsustainable levels. The balance that Rantideva found—through restraint, generosity, and living in harmony with nature—is the lesson for our times.'

Raghav, moved by the story and the sage's wisdom, asked, 'But how can we, who are so accustomed to indulgence, learn to live simply, to reduce our desires?'

Vasishtha smiled gently and said, 'The key lies in conscious living. Tapasyā is not about renouncing all joy, but rather about understanding the difference between needs and desires. It is about knowing that the more we consume, the more we take away from the Earth, from future generations, and from the balance of life. True fulfilment is not found in excess, but in mindful restraint—in choosing to live with what is necessary and living with intention.'

A Lesson for the Future

As Raghav left the sage's presence, the teachings of Tapasyā began to take root in his heart. The prince understood that the future depended on how humanity chose to live, how it balanced its desires with the limits of nature. He returned to his kingdom and led by example—adopting simpler ways of living, advocating for the conscious use of resources, and promoting generosity over consumption.

The story of King Rantideva and the principles of *Tapasyā* provide an enduring lesson for the future. In a world overwhelmed by consumption, the practice of self-restraint, conscious consumerism, and reducing unnecessary desires is not only a spiritual discipline but a necessity for sustainability. Just as King Rantideva's kingdom flourished through simplicity, so too can our world find balance by living with less, consuming mindfully, and respecting the delicate equilibrium of nature.

In the end, *Tapasyā* teaches us that the path to true happiness and peace lies not in indulgence, but in harmony—with ourselves, with nature, and with the world around us.

Ripples of Free Will: Beyond the Boundaries of Balance

In the timeless dance of existence, where human free will treads alongside cosmic forces, there lies a delicate balance—a harmony that must be nurtured and protected. It is this balance that underpins every action, every thought, and every choice that we, as humans, make. In the sacred stories we've journeyed through together, there lies an invitation to reflect, to question, and to align ourselves with a greater truth that transcends the ego-driven desires of the moment.

At the heart of these stories lies one central theme: the need to align our free will with the deeper currents of *Dharma*, spirituality, and the balance of nature. Our world is a living organism. It is interconnected, breathing in and out through the cycles of time. And just as a tree grows in harmony with its surroundings, each of us must grow, too. But to grow with wisdom, we must understand the role of our free will—not as a weapon to wield, but as a tool to align ourselves with the cosmic dance of creation.

Vasudhaiva Kuṭumbakam

The world is one family. This simple phrase carries within it the weight of centuries of wisdom, reminding us that the Earth, the rivers, the forests, and every living being, including ourselves, form one vast, interconnected family. It speaks of a time when the boundaries between 'us' and 'them' did not exist. There was no distinction between the human and the non-human, for every living being was seen as a reflection of the same divine essence.

Yet, as time moved forward, the complexities of life multiplied, and with it, our sense of separateness grew. In our pursuit of progress, we began to believe that we, the human race, were separate from the natural world, and that the Earth existed to serve our whims and desires. But the true essence of *Vasudhaiva Kuṭumbakam* calls us back to our roots. It whispers to us that the well-being of one is intimately tied to the well-being of all. This story serves as a reminder of the responsibility we have, to care for the Earth, to treat it as our family, and to make choices that consider the future of all its inhabitants. The world will only flourish if we remember that we are a part of it, not its masters.

Prakṛti Rakṣā

The protection of nature, is not a passive act. It is a deliberate choice—a vow to walk in harmony with the Earth. For centuries, ancient cultures have understood that every action we take creates ripples across time and space. When we destroy, we leave scars. When we nurture, we sow seeds of future growth. To protect nature is to protect ourselves, for in our stewardship of the land, the rivers, and the forests, we secure the very life force that sustains us.

But *Prakṛti Rakṣā* is not just about conserving the environment; it is about recognising that our very essence is intertwined with the Earth. It is a call to return to a way of life where every choice we make is guided by reverence for the natural world. This principle teaches us that our free will must be exercised with the awareness that we are part of a larger web of life, and every action has the power to strengthen or weaken that web.

In this world, our free will is often exercised in isolation, detached from the consequences of our actions. But the stories of the past, like those of the ancients who lived in deep communion with nature, beckon us to remember the interconnection of all things. If we want to see a future where our children inherit a world of abundance, then we must act now. We must protect nature, not as an obligation, but with love and devotion to the very land that gives us life.

Ahimsā

Non-violence carries with it an understanding that every living being has a sacred right to exist, and that to harm another is to harm the self. The true meaning of *Ahimsā* goes beyond mere physical non-violence; it calls us to cultivate compassion and mindfulness in every aspect of our lives—towards the Earth, towards animals, and towards each other. When we harm, we disrupt the natural harmony that exists between all beings.

In our modern world, *Ahimsā* is often misunderstood as a passive act, or mere avoidance of physical harm. But true *Ahimsā* is active—it is the commitment to live with kindness, to choose compassion over destruction, and to recognise the sacredness in all things. When we apply *Ahimsā* to our free will, we make choices that reflect a deep respect for life in all its forms. Our free will is not a license to dominate or destroy; it is a gift to nurture, to heal, and to protect.

Pancha Mahabhuta

The principle of the five elements, teaches us that everything in existence—our bodies, the Earth, and the cosmos—is made of the same fundamental elements: Earth, Water, Fire, Air, and Ether. These elements are not separate from us; they are within us, and we are within them. This truth reveals that the Earth is not something outside of us; it is a reflection of our own nature.

The five elements remind us of our responsibility to maintain balance within ourselves and with the world around us. Just as the elements must be in balance to create life, so too must our actions be balanced with the needs of the Earth. If we deplete the Earth's resources recklessly, we disrupt the balance of the elements, and in doing so, we disrupt our own balance. The story of the five elements urges us to listen deeply to the natural world, to honour its rhythms, and to understand that our free will must be exercised with reverence for the balance that sustains all life.

Tapasyā

Austerity and discipline, is perhaps one of the most challenging principles to embrace in the modern world. We live in an age of excess, where indulgence is often seen as a mark of success. *Tapasyā* teaches us the power of restraint—the ability to forgo unnecessary desires and live with intention. In an age where consumerism reigns supreme, *Tapasyā* offers us a path of liberation. It calls us to live simply, to consume mindfully, and to recognise that true wealth lies not in what we possess, but in what we give away.

When we embrace *Tapasyā*, we recognise that our free will is not meant to serve our fleeting desires, but to align with a greater purpose. It is through discipline that we can shed the layers of ego and reconnect with the deeper currents of existence. *Tapasyā* teaches us that true freedom comes not from indulging every whim, but from learning to live with balance, to recognise what is essential, and to let go of what is unnecessary.

Aranya

The wilderness teaches us that the Earth is not a resource to be exploited, but a sanctuary to be protected. The wilderness is not an untamed space but a reflection of the sacred balance of life. In the forests and mountains, we are reminded that life is not meant to be controlled; it is meant to be nurtured. The story of *Aranya* calls us to return to the wilderness, not to conquer it, but to learn from it.

In our modern world, we often look at nature as something separate from us, something to be tamed or subdued. But the story of Aranya reminds us that we are not separate from nature. We are a part of it. And when we harm the wilderness, we harm ourselves. The free will we exercise must be tempered with the wisdom of the wild. We must learn to live in harmony with nature, to protect the sacred places, and to return to the Earth's rhythms.

Dharmic Exchange

This teaches us that free will is not just about individual choice, but about collective responsibility. When Bali's ambition led him to challenge the cosmic order, he was reminded that true power comes from understanding our place in the world, not from trying to control it. The cosmic dance of free will reminds us that we are not alone in this universe; our choices affect the whole.

As we make our choices, we must remember that they ripple outward, affecting not just our lives, but the lives of all beings. The balance of the cosmos is fragile, and when we disrupt it, we disrupt the harmony of the world. The story of Bali calls us to reflect on our own ambitions—are they rooted in humility, or are they driven by ego? In a world where ambition often eclipses wisdom, it is a reminder that true greatness comes from understanding our place within the cosmic balance.

The Silence of the Bow

This story teaches us that there is strength in restraint. In a world that values action and speed, silence offers us the wisdom to pause, reflect, and act with precision. The bow, when drawn too tight, can snap. The mind, when filled with noise and distraction, can lose its way. Silence is the space where wisdom is born, and it is in this silence that we find the true power of free will.

Annam Brahma

'Food is Brahman' teaches us that the way we nourish ourselves is sacred. Food is not just sustenance for the body; it is a gift from the Earth. When we consume mindfully, we honour the Earth's abundance. We must recognise that our free will is not only exercised in our actions but in the way we care for our bodies. By treating food as sacred, we reconnect with the Earth's rhythms and remember that we are stewards of this planet, not its conquerors.

As we reflect on these stories, we come to realise that our free will is not a solitary force, cut off from the world around us. It is a thread woven into the very fabric of existence, one that connects us to everything—to each other, to the Earth, and to the cosmic rhythm that sustains life itself. We often think of free will as an individual privilege, an untamed power that allows us to act according to our desires. But the truth is, free will is a shared responsibility. Every choice we make sends ripples through the universe—ripples that affect not only the course of our lives but also the lives of those who come after us.

Imagine a single drop of water falling into a calm pond. At first, it seems insignificant, yet the ripples stretch far beyond the point of impact, touching every corner of the water's surface. Similarly, our decisions, no matter how small, extend far beyond our immediate environment. The way we treat others, the choices we make in how we interact with the Earth, the respect we show for life—it all matters. Each of these actions, rooted in free will, reverberates through time, shaping the world we will leave behind. This is where the delicate balance lies. Free will, when unchecked and acted upon impulsively, becomes a force of destruction—a powerful wave that crashes against the shores of wisdom and harmony. However, when free will is centred, when it is guided by wisdom, compassion, and reverence for all life, it becomes a force of creation. It has the power to heal, to restore, and to bring about a future where harmony reigns.

Take, for example, the act of planting a tree. In the moment, it might seem like a simple act—a seed placed in the Earth. But this act, rooted in conscious choice, creates a ripple that will extend far beyond the life of the tree itself. That tree will provide oxygen for years to come, shelter for countless creatures, shade for weary travellers, and beauty for all who gaze upon it. A single act of mindful free will, rooted in a deeper understanding of our interconnectedness, can transform the world around us.

The same principle applies to our daily choices—whether we decide to reduce waste, to act with kindness, to be present with our loved ones, or to protect the natural world. These are not just isolated actions; they are the small, intentional ripples that, when multiplied across time, create waves of change. Each action, no matter how small, is a step towards the balance we seek.

The stories shared here—of *Vasudhaiva Kuṭumbakam*,

Prakṛti Rakṣā, Ahimsā, and the others—are not mere myths or folklore. They are timeless truths, embodying the eternal dance between our free will and the wisdom of the Earth. They offer us a map—a guide to reconnect with the deeper currents of life, to align our will with the heart of the Earth, and to understand the sacredness of every action we take. But this journey is not one we can take lightly. It is a journey of remembering. We must remember the wisdom of the past—the stories of the sages, the lessons of nature, and the teachings of our ancestors.

We must remember that we are not separate from the Earth; we are of it. And as we walk this path, we must realign ourselves with the balance that sustains us. We must remember that every choice, however small, has the power to either restore or disrupt that balance. Our journey towards a balanced, harmonious world begins with each of us—right here, right now. The choices we make today will determine the world of tomorrow. The future is not some distant, abstract place; it is the culmination of every action we take in the present. If we are to shape a future of peace, harmony, and sustainability, we must begin with ourselves. We must choose to centre our free will, to act with intention and wisdom, and to remember that we are part of a greater whole.

So, let us tread carefully, with humility and reverence, for the future is not a far-off dream—it is being written in the choices we make right now. Each moment offers us the opportunity to realign with the sacred flow of life, to let our free will be guided by wisdom, and to create ripples of love, peace, and balance that will echo through generations. The world is waiting for us to remember the sacredness of our actions, to return to the heart of the Earth, and to shape a future where harmony is not just an ideal, but a living, breathing reality.

From Vedas to Vision 2070

India is blessed with everything that any country can hope for. Sprawling geography that covers all terrains, with ample availability of sunlight, balanced climate throughout the country, where all four seasons are evenly spread, a billion-plus hearts, and the massive talent pool that comes with it, a growth potential, and the way the country is transforming itself with ambitious milestones. The whisper from the sun-bathed Indian subcontinent is pertinent. That whisper carries with it the echoes of ancient hymns, the wisdom of sages, and the dreams of a billion hearts. This is the voice of India, a land where sustainability is not a policy, but a legacy deeply ingrained in its cultural fabric. With the country's recent commitment to achieve net-zero emissions by 2070, the nation commences to embark on a massive transformation journey that stitches together tradition and innovation.

This section will take you through the diverse and dynamic initiatives that define India's sustainability roadmap. We explore five elements, *Surya* (the sun), *Vāyu* (air), *Jal* (water), *Agni* (fire) and *Prithvi* (Earth) create renewable energy, promote efficient systems, and preserve ecological balance. Each element is explored through a lens of cultural reverence and modern advancement.

Space (Ākāśa), has been deliberately left out of the following discussion because it incorporates and simultaneously exists in all the other elements while itself being formless. Agni (Fire), on the other hand, is discussed in two forms; as Āditya (Sun) and as Agni (Fire).

As a growing nation, the type of growth we need is starkly different from what the western countries need. Their growth is dependent on a major switch or upgrade of existing systems, while India needs to decarbonise growth itself. Every section of society needs to be empowered to adopt greener practices and improve efficiency. From revitalising sacred rivers to mirroring the cycle of *Samsara* and adopting modern-day sustainable financial practices, India can be a torchbearer for equitable and sustainable development. As you learn more about India, expect to uncover a harmonious blend of scientific rigour and spiritual insights, giving you a glimpse into the immense work being done to balance progress with preservation. This is not merely a story of policies and plans; it is a clearly mapped road demonstrating India's commitment to safeguard the planet for generations to come.

The Five Elements

Āditya (Sun)

In the Rigveda, the sun is celebrated as *Āditya*, the life-giver, the source of energy and enlightenment. Drawing inspiration from this reverence, Bharat, launched the National Solar

Mission, aiming to harness the power of the sun to achieve 500 GW of renewable energy capacity by 2030. This mission, supported by the International Solar Alliance, is in absolute resonance with *Vasudhaiva Kuṭumbakam*—the world is one family. Through collective action, India is transforming sunlight into a universal resource, lighting up homes, and hope across the globe.

Underneath the blazing sun, villages are being adorned with solar panels that glisten like sacred offerings. Imagine the year is 2070 and you look back into the past on what India has done. What you would see is immense transformation that impacts every life. From the rugged terrain of Ladakh, where solar installations provide electricity to the most remote villages, to the bustling streets of Delhi, where solar energy powers a vibrant economy, the nation's energy transition is both deep and wide reaching. Backed by microgrids, rural area energy is completely transformed, energy is self-sufficient and domestic power consumption is literally free to the entire country. The shift to solar energy has not been limited to just electricity generation, but has permeated all sectors from transportation to agriculture.

Solar-powered EVs traverse the highways, reducing reliance on fossil fuels while contributing to cleaner air. Agricultural innovations, such as solar irrigation systems, have transformed farming practices, ensuring that crops thrive even in drought-prone areas. Solar energy has provided farmers with the tools to enhance productivity, reduce costs, and improve their livelihoods. In this future, no longer does the sun's energy serve as an untapped potential; it is the pulse of the nation's growth and the foundation of its future.

The cities of India, once known for their traffic-choked

streets and rising pollution levels, have transformed into smart, sustainable urban hubs. Powered by solar energy, these cities are built on green principles, with buildings designed to be energy-efficient and equipped with solar panels that generate more energy than they consume.

Public transportation systems, which were once a major source of pollution, are now electric, reducing their carbon footprint and enhancing the quality of life for urban dwellers. In 2070, India's journey toward solar energy has been much more than a technical achievement. It is a testament to the nation's commitment to Dharma—the moral responsibility to protect the Earth and its people.

The story of India's solar revolution is not just about panels and power plants; it is about the people's unwavering belief in the sun's power to unite, to uplift, and to transform. It is about harnessing the energy of the universe and using it to elevate the lives of all—whether in the cities or in the quiet remote corners of the country. The solar revolution played a crucial role in combating climate change.

By investing in renewable energy and leading the global effort to reduce carbon emissions, India has helped to steer the world away from the brink of environmental catastrophe. Through collaboration, innovation, and collective will, the sun's energy has become a beacon of hope, lighting the way to a sustainable future for the planet.

As India moves towards its 2070 vision, the legacy of the National Solar Mission will be felt not just in the number of gigawatts generated or the rise in clean energy jobs, but in the transformation of the Earth itself—a planet that, in the words of the *Rigveda*, is *sustainably nurtured and eternally revered*. In this future, Earth and humanity are no longer at odds. Instead, they have found harmony through renewable energy, guided by the eternal wisdom of *Vasudhaiva Kuṭumbakam*—the realisation that all of us are connected, and by protecting the Earth, we protect ourselves and each other. The sun, as it always has, continues to shine, but now, it does so with a purpose, lighting up not just the world, but the path to a brighter, more sustainable future for all.

***Vāyu* (Air)**

Air is not a physical element but a symbol of life, movement, and vitality. It is through Vāyu that life flows and evolves, unseen, yet ever-present, connecting all beings in a web of breath and energy. This elemental force has inspired India's forward-looking vision through the National Hydrogen Mission, marking a transformative step into the future. By producing green hydrogen, an eco-friendly fuel, harnessed from the abundant natural resources of wind and water, India is embracing the invisible energy of *Vāyu* in its purest form.

Much like the teachings of the Upanishads, which highlight the interconnectedness and balance of all elements, the National Hydrogen Mission embodies the idea of interdependence. Just as air cannot be owned but is shared by all, green hydrogen is intended to be a resource for the collective good, a fuel that nurtures not just one, but the entire planet. It is an energy source that promises to transcend borders and systems, aiming for sustainability and universal accessibility, embodying the principle of *Vasudhaiva Kuṭumbakam*.

In 2070, the vision of green hydrogen production will have evolved into a cornerstone of India's sustainable energy landscape. Picture a sprawling industrial complex in

Gujarat, once known for its heavy pollution and emissions, now transformed into a centre for green hydrogen production. The chimneys that once billowed dark smoke now release the clean, gentle steam of a revolutionary process—green hydrogen that powers industries and homes alike, with minimal environmental impact. It is a transformation of both technology and consciousness, where the old methods of pollution and waste give way to innovation and sustainability.

By 2070, India's National Hydrogen Mission will have fulfilled its promise, not only having met domestic energy needs, but also having positioned India as a global leader in green hydrogen technology. From rural farms in Rajasthan, to urban centres in Bengaluru, green hydrogen will fuel transportation, industry, and power systems, creating a decentralised, resilient energy infrastructure that relies on natural elements—wind and water—that sustain life on Earth.

The production of green hydrogen, like *Vāyu*, will be woven into the fabric of daily life. Just as air breathes life into all living beings, hydrogen will energise industries and homes without the environmental costs associated with fossil fuels. India's leadership in green hydrogen will go beyond economic and technological power; it will embody a vision of a world where nature and progress move forward in harmony, where innovation is aligned with the well-being of the planet.

In this future, the air, once taken for granted, will no longer be seen as an invisible resource to be polluted but as a shared force that powers civilisations. Green hydrogen, as an energy source born of *Vāyu*, will unite humanity in a collective pursuit of sustainability, transcending national boundaries, uniting people in the shared responsibility to care for the Earth and all its inhabitants.

The National Hydrogen Mission, by harnessing the invisible power of nature and grounding it in the ancient wisdom of balance and interdependence, will have not only reshaped India's energy future but will have helped guide the world to a cleaner and brighter tomorrow. The future of energy, like the wind, will be shared, flowing freely across nations, and sustaining life for generations to come.

Jal (Water)

Water is revered not merely as a resource, but as a sacred force of purification, sustenance, and life. It is the essence of life itself, the giver of fertility and the sustainer of all that grows. The Upanishads speak of water as one of the five elements, essential to the harmony and balance of the universe. From the sacred rivers like the Ganga and the Yamuna, to the rain-filled clouds that nurture the Earth, *Jal* symbolises both physical and spiritual nourishment. It is, therefore no surprise that India's commitment to water conservation has deep roots in this reverence and the Jal Shakti Abhiyan launched to ensure sustainable water use, echoes this ancient wisdom. The Jal Shakti Abhiyan is India's modern-day call to action, drawing upon the ancient systems of water management that have long sustained the land.

For centuries, communities across India have devised ingenious ways to collect, store, and manage water. From the stepwells of Gujarat, to the tanks of Tamil Nadu, traditional water management systems were built with an understanding of *Jal's* cyclical nature. These structures served as vital lifelines. The revival of these traditional methods, integrated with modern technologies, like watershed management and rainwater harvesting, is a celebration of the wisdom of the past and innovations of the future.

By 2070, the vision for water management in India will have evolved into a comprehensive, sustainable model that blends traditional knowledge with cutting-edge technologies. Imagine a future where India's landscapes—urban and rural—are dotted with decentralised, smart water management systems that work in harmony with nature. From the deserts of Rajasthan to the plains of Uttar Pradesh, water conservation will be the cornerstone of India's climate resilience strategy. The nation will have transformed itself into a global leader in water sustainability, leveraging both innovation and reverence for nature to ensure water security for all.

Rajasthan, once known for its arid landscapes and recurring droughts, will be a beacon for sustainable water use. Traditional *johads*, earthen check dams that have stood the test of time, will have been revived and modernised. These *johads*, combined with advanced rainwater harvesting techniques and artificial recharge systems, will provide sustainable solutions to water scarcity.

In these arid regions, the fusion of ancient wisdom and modern engineering will ensure that every drop of rainwater is captured, stored, and efficiently used. As a result, groundwater levels will rise, irrigation will be more efficient, and the once-dry land will flourish with agriculture, supporting not just local communities, but also contributing to national food security.

In rural areas, where water scarcity has historically posed

significant challenges, India will have pioneered sustainable, community-driven water management initiatives. Farmers will rely on advanced drip irrigation systems powered by solar energy, ensuring that every drop of water is used to nourish crops without wastage. In many parts of the country, rainwater harvesting systems will be commonplace, allowing villages to be self-sufficient in managing their water needs. The practice of building community-managed water bodies, once a hallmark of rural India, will be revived, fostering a spirit of collective responsibility and ensuring that water resources are equitably shared.

India's water management model will also have made significant strides in addressing the challenges posed by climate change. With unpredictable rainfall patterns and changing monsoon cycles, India will have invested in cutting-edge technologies like cloud seeding, desalination, and water recycling to ensure a steady and reliable water supply. India's rivers, once considered over-exploited, will be rejuvenated through large-scale river basin management projects that prioritise ecological balance and sustainable use.

Jal will once again flow freely, nourishing the land, and sustaining life as it always has. This transformation in India's relationship with Jal will be more than a technological or infrastructural achievement. It will be a profound shift in the nation's ethos, rooted in a deep respect for water as both a physical and spiritual force. The act of conserving and managing water will be seen not just as a practical necessity but as an act of reverence—an acknowledgment of Bhūmi Devi.

The vision for 2070 is one where water is no longer a commodity to be fought over, but a shared resource to be cherished and protected. In this future, *Jal*—like *Āditya* and *Vāyu*—will continue to be a source of life, a binding force that connects all living beings, and a key element in India's sustainable growth. Through the Jal Shakti Abhiyan and its commitment to water conservation, India will ensure that the sacred waters of the Earth are preserved for future generations, fostering a world where sustainability, interdependence, and reverence for nature are the guiding principles of all progress.

Agni (Fire)

Fire holds a sacred place in Indian culture, as the symbol of transformation, purification, and energy. It is the force that ignites creation, the energy that fuels life, and the transformative power that purifies both body and spirit. In a *yajna*, Agni represents not just destruction, but more importantly, regeneration and renewal.

It is this timeless understanding of fire's dual nature that India harnesses today in its drive for energy efficiency and clean energy innovations. The modern expression of Agni in India is reflected in its commitment to energy efficiency and sustainable growth.

Programmes such as the Ujala Scheme, launched to promote the widespread adoption of LED lighting, have fundamentally changed the way energy is consumed across millions of households.

By shifting from traditional incandescent bulbs to energy-efficient LEDs, India has not only reduced energy consumption but has also significantly cut down on carbon emissions. This simple shift mirrors the ancient reverence for fire's capacity to illuminate, cleanse, and transform. Just as the fire in a *yajna* purifies offerings, LED lighting purifies the energy grid, making it cleaner and more sustainable for future generations.

By 2070, the vision for *Agni* will have expanded beyond household lighting into a full-fledged transformation of the country's energy landscape. India will have pioneered a new era of energy efficiency, where every home, every business, and every factory is optimised for minimal energy waste and is powered by renewable energy grids. The rise of advanced battery systems, coupled with smart grids, will ensure that energy, especially from renewable sources like solar and wind, is stored and distributed with maximum efficiency.

These smart grids, much like the controlled use of fire in sacred rituals, will regulate the flow of energy, ensuring that it is used purposefully and without waste. Through this innovation, India will not only reduce its reliance on fossil fuels but will also contribute to the global effort to limit climate change and protect the planet.

In 2070, India's energy future will be deeply intertwined with cutting-edge technologies that continue to push the boundaries of efficiency and sustainability. Picture a bustling city like Bengaluru, where startups lead the way in developing groundbreaking energy storage solutions. These innovations will make it possible for India to harness the full potential of renewable energy, storing excess power generated during the day and ensuring a steady, reliable supply even when the sun sets or the wind dies down.

Battery storage will have become a critical pillar of India's clean energy infrastructure, providing the flexibility needed to transition fully to a renewable energy future. These energy storage systems will be as vital to the country as the controlled, purposeful flames of *Agni*, which sustain life and drive progress.

India's energy landscape will also see a revolution in how industries consume and conserve energy. Smart manufacturing processes, powered by artificial intelligence and machine learning, will enable factories and production plants to optimise energy use, reduce waste, and increase efficiency. Just as fire in ancient times was used to forge tools and create new materials, India will use the transformative power of energy-efficient technologies to forge a new path of industrial growth—one that is sustainable and environmentally responsible.

The power of *Agni* in India's energy strategy will also be evident in the rise of green technologies. From EVs to green hydrogen production, India's innovation in clean energy will redefine how the world sees energy consumption. EVs, powered by renewable energy, will have replaced internal combustion engines on the roads, dramatically reducing air pollution and contributing to the reduction of greenhouse gas emissions.

In the field of green hydrogen, India will lead the way in producing hydrogen fuel from renewable sources, a clean alternative to fossil fuels that will power industries, transportation, and even homes, creating a zero-emissions energy ecosystem.

India's mastery of *Agni* will be a case to the country's commitment to sustainable development and for other countries to emulate. The destructive potential of fire, once feared and controlled, will have been harnessed and redirected into a force for positive transformation. This transformation will not only impact India's energy landscape but will have a ripple effect across the globe. As India becomes a leader in energy efficiency and clean energy technologies, the world will look to its innovations as a model for sustainable growth.

The integration of energy-efficient technologies with India's traditional reverence for fire, *Agni*, will also redefine how the nation interacts with nature. India will be a leader in adopting energy systems that are both technologically advanced and ecologically responsible, ensuring that progress does not come at the cost of the environment. Through its innovations, India will continue to honour the teachings of *Agni*, recognising that transformation and purity are achieved not through destruction, but through balance and conscious action.

The future of energy in India, powered by the strength of *Agni*, will be one of purposeful transformation—where every spark of innovation contributes to a cleaner, greener, and more sustainable world. It will be a future where fire no longer symbolises unchecked consumption, but rather the regenerative power of innovation, leading the world into an era of energy efficiency and sustainability. Just as the flames of a sacred fire purify offerings, India's mastery of *Agni* will purify its energy systems, igniting a brighter, more sustainable future for generations to come.

***Prithvi* (Earth)**

Earth is revered as the bedrock of life, stability, and sustenance. She is the nurturing force that supports all living beings, providing a foundation for growth, harmony, and balance. In Vedic hymns, *Prithvi* is described as the mother who holds the cosmos together, a steady, unshakeable presence that sustains life and nurtures creation.

This reverence for *Prithvi* extends beyond spiritual practices into India's modern-day efforts to build a resilient, sustainable future. The principle of *Prithvi* underscores India's commitment to creating climate-resilient infrastructure and promoting a circular economy that reduces waste, conserves resources, and enhances the well-being of both people and the planet. India's efforts to strengthen its climate resilience are a direct response to the challenges posed by climate change, urbanisation, and environmental degradation.

The rise of circular economy principles—where waste is minimised, and products are reused and recycled—aligns seamlessly with the wisdom of *Prithvi*. Just as the Earth recycles nutrients, minerals, and organic matter to maintain the delicate balance of ecosystems, India is pioneering ways to reuse materials and reduce the environmental impact of its development. The commitment to building resilient infrastructure using recycled materials is an acknowledgment of the need to protect *Prithvi*, the foundation that sustains all life, and to honour the Earth's ability to regenerate and sustain us.

By 2070, India's commitment to *Prithvi* will have reshaped its approach to construction, urbanisation, and environmental management. Picture a bustling city like Mumbai or Pune, where green buildings made from sustainable materials, are energy-efficient but are also made from recycled materials, reducing the need for new raw materials and minimising the environmental footprint of construction.

Concrete will be replaced by recycled aggregates, steel by reclaimed metal, and plastic waste will be transformed into innovative building components. The construction industry, once a major source of waste and resource depletion, will now lead the way in sustainable practices that respects Earth's stability and resilience.

In rural areas, India's focus on sustainable infrastructure will be no less transformative. In Maharashtra, for example, an innovative housing project that uses construction materials derived from industrial waste will have become a model for sustainable living. The use of recycled materials, such as fly ash, plastic waste, and discarded metals, will not only reduce the pressure on landfills but also contribute to the creation of low-cost, energy-efficient homes. These homes, designed to withstand the effects of extreme weather events, will stand as symbols of India's resilience in the face of climate change. The materials used in their construction will not only reduce the environmental impact but will also be locally sourced, providing jobs and economic opportunities for communities that have long struggled with poverty.

India's vision for 2070 also includes the widespread adoption of sustainable practices in agriculture and waste management. In rural landscapes, where Prithvi is the primary sustainer of life, farmers will use organic fertilisers and biodegradable products that regenerate the soil and enhance its fertility. The circular economy model will extend into waste management systems, where organic waste is composted, and recyclable materials are processed and reused.

The drive for zero waste will be a cornerstone of India's sustainability strategy, with communities across the country adopting practices that reduce waste and increase the recycling of resources.

By 2070, India's cities and rural areas will have undergone a complete transformation, with both being resilient, self-sustaining ecosystems that thrive on principles of balance, regeneration, and respect for *Prithvi.* Green infrastructure—such as urban forests, green rooftops, and water-permeable surfaces—will have been integrated into city planning, helping to reduce heat islands, improve air quality, and promote biodiversity. These cities will not only be energy-efficient and water-sustainable, but they will also serve as living examples of how humanity can coexist harmoniously with nature, building on the principles of *Prithvi* to create urban environments that are both resilient and restorative.

India's commitment to *Prithvi* will also manifest in its efforts to protect and restore ecosystems. Large-scale reforestation programmes, conservation of natural habitats, and the revival of rivers and wetlands will have become central to the nation's environmental strategy. The recognition that the Earth is not merely a resource to be exploited but is a living, breathing entity to be protected and revered will be a guiding principle of all of India's environmental policies.

By 2070, the country will have succeeded in reversing much of the environmental degradation that has occurred over the past century, ensuring that *Prithvi* remains a stable, nurturing force for future generations. As India approaches its 2070 vision, its efforts to build climate-resilient infrastructure and promote a circular economy will be seen as a testament to the nation's deep reverence for *Prithvi.* These efforts will be more than technological or industrial achievements—they will reflect a profound cultural shift towards sustainability—a shift grounded in the ancient wisdom that the Earth is both a gift and a responsibility.

India's future will be one where the Earth's stability is not just maintained, but cherished, and where the balance between progress and nature is honoured in every aspect of life. In this future, Prithvi will continue to be the foundation of life, a stable, enduring force that nurtures all life and supports humanity's journey toward a sustainable, harmonious world.

To truly understand the journey of decarbonising India's growth, it is essential to focus on the present—our current struggles, challenges, and opportunities. India, in its quest for industrialisation and economic progress, has been one of the fastest-growing economies in the world. However, this rapid growth has also been accompanied by an increasing carbon footprint, driven largely by energy-intensive industries, urban expansion, and the rising demand for transportation.

The challenge before India today is not just to continue its growth, but to do so in a way that is sustainable—balancing development with the urgent need to combat climate change. In essence, India's current efforts to decarbonise its growth reflect an ancient principle deeply embedded in the fabric of Indian culture and wisdom: *Dharma*.

In the Mahabharata, *Dharma* is often portrayed as the righteous path—an ethical and moral framework that guides individuals and nations alike in their actions. It is not a rigid set of rules, but a dynamic principle that adapts to context and time, always urging us toward balance, fairness, and responsibility. The essence of *Dharma* is deeply connected to the natural world and the harmony that must exist between humans and the environment.

The present focus of India's decarbonisation efforts is anchored in policies and initiatives designed to reduce carbon emissions and promote sustainable practices across sectors. The Perform, Achieve, and Trade (PAT) Scheme, for instance, encourages industries to reduce their specific energy consumption and carbon emissions. It's a system that rewards industries for energy efficiency, driving them toward more sustainable practices.

In industrial hubs like Surat and Pune, factories are embracing energy-efficient technologies such as waste heat recovery systems and transitioning to renewable energy sources like solar and wind. These changes are not just about reducing emissions; they are about making businesses more cost-efficient, more competitive, and more responsible in the long run.

Just as *Dharma* calls for a balanced approach to life—where righteousness is aligned with practical wisdom—India's industries are finding that reducing their carbon footprint also results in long-term financial benefits. This reflects a modern interpretation of *Dharma*: aligning progress with sustainability.

Public transportation systems are being electrified, with metro networks and e-buses reducing the carbon footprint of daily commuting. These changes reflect a growing recognition of the interdependence between urban development and environmental responsibility.

Just as Dharma emphasises the interconnectedness of all living beings and the world around us, India's urbanisation is now evolving to consider the impact on the environment. The cities of today are learning to harmonise progress with ecological sensitivity, proving that urbanisation and sustainability can coexist.

One of the key drivers of this transformation is the focus on small and medium enterprises (SMEs), which form the backbone of India's economy. These businesses often struggle to access the capital and technology needed to adopt greener practices. However, India is now offering financial incentives, support, and technology transfers to help SMEs transition to cleaner, more energy-efficient models.

This approach mirrors the ancient principle of *Dharma*, which calls for inclusivity and fairness—ensuring that in all sectors, even the smaller players are part of the collective effort. The transition to a low-carbon economy is not only the responsibility of large corporations or governments, but it must be a shared endeavour, with every segment of society contributing to the broader goal of decarbonisation.

However, the path to decarbonising India's growth is not without its challenges. The transition requires significant investment, technology development, and policy coordination. The struggle to reduce carbon emissions and promote sustainable growth is not a short-term battle, but a long-term commitment.

Looking to the future, India's commitment to decarbonising its growth will set a precedent for the world. As the largest democracy and one of the fastest-growing economies, India has the potential to lead by example, showing that growth and sustainability are not mutually exclusive. Balanced action, responsibility, and fairness will guide India in its transition to a sustainable, low-carbon economy. By making energy-efficient and environmentally responsible decisions today, India is laying the foundation for a better future, not only for its citizens but for the world at large.

Just as the Mahabharata teaches us that the righteous path leads to victory, India's path of decarbonisation will lead to a future where growth, prosperity, and environmental sustainability are aligned, and where the wisdom of the past guides the actions of the future.

Satat Krishi (Sustainable Agriculture)

Sustainable Agriculture emphasises farming practices that meet current food needs, without compromising the ability of future generations to meet their own needs. Sustainable agriculture integrates ecological balance, economic viability, and social responsibility with a focus on long-term environmental health and productivity.

In the *Atharva Veda*, Earth is worshipped as a life giver, the sustainer of all beings. This ancient relationship between humans and the land is reflected in the country's growing shift towards *Satat Krishi*. The rise of agroforestry is particularly significant. In agroforestry systems, farmers plant a mix of crops alongside trees—fruit trees, nitrogen-fixing trees, and timber species. This not only provides multiple sources of income but also improves soil health by preventing erosion, enhancing water retention, and promoting biodiversity.

Agroforestry, which echoes ancient practices where forests and agricultural lands coexisted symbiotically, helps farmers create more resilient farming systems. In the heart of Madhya Pradesh, a young farmer has embraced agroforestry, blending crops like wheat and legumes with fruit trees like mango and guava. His fields, once solely focused on monoculture crops, have now transformed into thriving ecosystems, fostering diverse plant and animal life, and reinforcing the deep bond between man and nature.

By 2070, India's agricultural system will have fully embraced sustainable practices that prioritise ecological balance, resource conservation, and resilience to climate change. These future farming systems will not only produce food but also act as guardians of the environment, protecting Bhūmi Devi from the depletion caused by overexploitation.

By 2070, *Satat Krishi* will have become the norm across India. Organic farming will no longer be a niche practice but a widespread movement that touches all corners of the country. The Indian government's support for organic certification programmes, along with consumer awareness of the benefits of organic food, will drive this change. Farmers will adopt regenerative agriculture, a set of practices that restore soil health, increase carbon sequestration, and promote biodiversity. Crop rotations, agroecology, and the use of cover crops will become standard practice, allowing farmers to restore the land's fertility and ensure that the soil remains productive for generations to come. Agroforestry will be fully integrated into India's rural economy by 2070.

In this vision, every farmer—whether in the plains of Uttar Pradesh or in the hilly terrains of Uttarakhand—will plant trees alongside their crops, creating multi-layered agroecosystems that provide food, fodder, timber, and medicinal plants. These systems will offer farmers the tools to diversify their income and reduce their reliance on seasonal crops, while simultaneously contributing to forest cover, which in turn mitigates climate change.

Technological advancements in precision agriculture will play a key role in sustainable farming by 2070. Farmers will use data-driven technologies, such as drones, sensors, and artificial intelligence, to monitor soil health, optimise water use, and reduce chemical inputs. Smart irrigation systems will ensure that water is used efficiently, conserving this precious resource in drought-prone areas. These innovations will help farmers make informed decisions, minimising waste and maximising crop yields, while maintaining ecological balance.

By 2070, India's agricultural systems will have fully embraced the principles of a circular economy. This will include recycling agricultural waste into compost, biofuels, and fertilisers, minimising waste, and reducing the need for external chemical inputs. Farmers will have access to advanced composting technologies, and organic waste from urban centres will be repurposed for agricultural use, creating a sustainable loop that benefits both urban and rural areas.

As climate change intensifies, India's agricultural systems will evolve to focus on climate-resilient crops and practices. The use of drought-tolerant and heat-resistant varieties of crops will become widespread, ensuring food security even in the face of changing rainfall patterns. Farmers will employ water-conserving techniques, such as rainwater harvesting and watershed management to adapt to unpredictable weather patterns. These efforts will build resilience against the adverse effects of climate change, ensuring that agricultural practices continue to thrive despite environmental challenges. The transition to sustainable agriculture will also involve empowering farmers with knowledge and resources.

By 2070, farming communities will be well-versed in the principles of *Satat Krishi*, as Bhūmi Devi, will be equipped with the tools and techniques needed to carry forward the tradition of sustainable farming. This transformation will be rooted in the belief that agriculture, like Bhūmi Devi, must remain nurturing and balanced, ensuring that the Earth's resources are preserved for future generations. Just as the Earth herself regenerates and sustains life, so too will India's agricultural systems thrive—abundant, resilient, and deeply connected to the land.

The Cycle of Samsara

This refers to the eternal cycle of birth, death, and rebirth. A continuous cycle of transformation that mirrors that natural world's rhythm of creation, preservation, and destruction. *Samsara* is not a static process; it is ever-evolving and ever-present, guiding life through endless cycles of regeneration. Just as each being is reborn into the world, the materials of this Earth too undergo cycles of creation, destruction, and renewal. This eternal cycle finds its modern reflection in the concept of the circular economy where resources are not wasted, but instead, they flow continuously through a system of reuse, recycling, and regeneration.

In both *Samsara* and circular economy, there is an inherent understanding that nothing is truly ever lost, everything is transformed, returning to the source in new forms. This cyclical view of the world, where destruction makes way for creation, is perpetually renewed and forms the foundation of India's evolving approach to sustainability.

India's journey into circular economy draws heavily from the spiritual and cultural values of *Samsara*, where the act of reusing, regenerating, and recycling becomes not just a matter of environmental responsibility, but a reflection of a deeper, spiritual truth: that the world is interconnected, and our actions ripple through the web of life.

Just as *Samsara* teaches that life and death are not opposites, but parts of the same cycle, circular economy views waste and resources not as dichotomies, but as interconnected elements in a continuous loop. India's push for a circular economy is not simply about technological or industrial advancements; it is a return to ancient wisdom that connects humanity to the Earth in a profound way—ensuring that the resources of today do not deplete the world of tomorrow.

As India steps toward 2070, the concept of a circular economy will be central to its strategy for sustainable growth. By 2070, the country will have fully integrated circular economy principles across industries, reducing the linear 'take-make-dispose' model and shifting toward a system where products are designed for longevity, reuse, and regeneration.

In this future, the concept of *Samsara* will be woven into the fabric of everyday life. Materials like plastic, metal, and even food waste will no longer be viewed as waste but as resources to be reused, repurposed, and reintegrated into the economy. Packaging will no longer be discarded after single use but will be made of biodegradable or recyclable materials, designed with *Samsara* in mind—where everything has a purpose and a future beyond its initial form.

Imagine a future in India where the waste from one industry becomes the raw material for another. In urban centres like Delhi and Bengaluru, smart waste management systems will be commonplace, where households and businesses separate their waste into categories such as organic, recyclable, and reusable. These waste streams will then be sent to specialised centres where they will be processed, purified, and reincorporated into the production cycle.

Industrial by-products, instead of polluting the environment, will be transformed into new materials, just as a soul undergoes a transformation through the cycles of *Samsara*. In this vision, India will embrace *Samsara* not as a concept of eternal suffering, but as a path to renewal and regeneration, ensuring that its economy runs not on exploitation, but on the virtuous cycle of sustainability.

By 2070, India's agricultural systems will also embody the principles of *Samsara*. In *Satat Krishi Samsara*, where every lifecycle feeds into the next, the agricultural system will operate in a closed-loop cycle, where nothing is wasted, and everything is reborn. This approach will not only preserve soil health but also reduce carbon emissions and ensure food security in the face of climate change.

In this future, India will also be a global leader in green technologies that reflect the teachings of *Samsara*. Waste-to-energy plants, for instance, will harness the energy of what is discarded, transforming waste into a valuable resource—much like how spiritual practices transform the ego and attachments into spiritual growth.

E-waste, rather than being dumped in landfills, will be carefully dismantled, with precious metals and components recovered and reused, feeding back into the production cycle.

EVs, powered by clean energy, will replace fossil fuel-dependent cars, reduce pollution and carbon footprints, and reflect the karmic principle that good actions—such as using renewable energy—lead to positive, regenerative outcomes for society.

At the heart of this transformation will be a profound cultural shift, driven by education and public participation.

By 2070, India's citizens will have fully embraced a circular economy, seeing it not just as a system for resource management, but as an extension of their spiritual and ethical values.

The principles of *Samsara*—rebirth, interconnectedness, and renewal—will guide the actions of individuals and communities. Waste will no longer be viewed as an undesirable by-product but as an opportunity for transformation, both for the individual and the community. Just as *Samsara* teaches the cyclical nature of life and death, India will teach that nothing is ever truly wasted—everything has a purpose, and everything can be transformed, renewed, and returned to the Earth.

In the India of the future, the vision of a circular economy will not just be an economic strategy but a spiritual practice—a return to the ancient wisdom that the Earth is our mother, and that it is our duty to care for her. By integrating the principles of *Samsara* with a circular economy, India will create a world where growth does not come at the expense of the Earth, where resources are respected and conserved, and where every cycle of production and consumption feeds into the next, ensuring a sustainable and harmonious future.

The circular economy will not just be an industrial or environmental strategy; it will be a path to spiritual and material balance, guiding India toward a more sustainable, connected, and regenerative future for all.

A Future Rooted in Tradition

As India marches towards net-zero by 2070, it does so with one foot in the past and the other in the future. This journey is not merely about reducing emissions but about rekindling a relationship with nature that has defined Indian civilisation for millennia. It is a testament to the enduring wisdom of the land, where progress and tradition coexist in harmony.

In the whispers of the wind, the flow of rivers, and the glow of the sun, India hears its ancient promise: to protect, nurture, and live in harmony with the Earth. This is not just a national journey; it is a message to the world, a reminder that sustainability is not a destination but a way of life through these twenty-five basic actions, which are in complete alignment with what our scriptures say.

- **Implement national circular economy policies**
  As the wheel turns, so must our actions return to the source, ensuring balance.–The Bhagavad Gita
- **Invest in green energy infrastructure**
  The sun's rays shine for all; harness them for the collective good. –Rigveda
- **Promote sustainable farming practices**
  Just as a farmer sows, so must we tend to the Earth with care and reverence. – Atharva Veda
- **Reduce single-use plastics**
  The ocean does not hoard the river's flow; let us not hoard that which harms.–Mahabharata
- **Embrace urban green infrastructure**
  Like trees, cities must grow in harmony with the Earth, offering shade and sustenance.–Manu Smriti
- **Facilitate waste segregation at source**
  Orderliness in the mind reflects in the world; let the waste be sorted with wisdom.–Yoga Sutras of Patanjali
- **Support eco-friendly businesses**
  Do not seek wealth at the expense of nature; let your profits nourish the land.–Kautilya's Arthaśāstra

- **Adopt energy-efficient technologies**
  The fire burns with purpose; let our energy be directed with intention.–Vedas
- **Encourage public transport over private vehicles**
  The many steps lead to a common goal; let not the journey be made alone.–Ramayana
- **Implement green building standards**
  Like the roots of a tree that anchor the forest, let our foundations rest upon the Earth's wisdom.–Mahabharata
- **Provide financial incentives for renewable energy adoption**
  The rain falls for all; let us share the blessings of abundance. –Rigveda
- **Reforest and restore natural habitats**
  The forest's breath is the Earth's heartbeat; restore it, and life shall flourish.–Bhagavad Gita
- **Educate citizens on sustainability practices**
  Knowledge passed down is the greatest gift; let it bloom in the hearts of the people.–Upanishads
- **Develop waste-to-energy technologies**
  From the ashes, the phoenix rises; let waste be transformed into light.–Puranas
- **Promote water conservation and management**
  The rivers must flow; the Earth must drink. Let us preserve the sacred stream.–Atharva Veda
- **Encourage minimalist and conscious consumption**
  Desire less, for what the soul craves is already within.–Bhagavad Gita
- **Support indigenous and traditional farming systems**
  The wisdom of the ancients is the soil from which the future grows.–Vedas

- **Invest in carbon capture and storage solutions**
  The air we breathe is our shared responsibility; cleanse it, and it shall nourish all. Yoga Sutras of Patanjali
- **Mandate sustainability in corporate governance**
  The tree grows tall through roots that reach deep; let governance grow in righteousness.–Mahabharata
- **Foster innovation in sustainable technologies**
  Creativity is the flame; let it burn bright to light the world with wisdom.–Bhagavad Gita
- **Implement sustainable consumption in the textile industry**
  Clothing made from the Earth must return to it with honour.–Manu Smriti
- **Increase forest and wildlife conservation effort**
  The jungle is not just a place; it is the breath of the Earth itself. –Bhagavad Gita
- **Develop efficient recycling systems for e-waste**
  What is cast aside today may serve its purpose tomorrow—restore it with respect.–Puranas
- **Promote local and sustainable food systems**
  The grains that grow on the land must nourish the hands that tend it.–Ramayana
- **Cultivate a cultural shift towards sustainability**
  When the heart is aligned with the Earth, all actions follow the path of righteousness.–Bhagavad Gita

Are We Listening?

Close your eyes and let the world fall silent for a moment. In the stillness, lean in and listen—not with your ears, but with your heart. Beneath the hum of machines, the rush of busy streets, and the clamour of human voices, there is another sound—a whisper. It has always been there, soft yet persistent, a voice carried by the rustle of leaves, the sigh of rivers, and the quiet breath of the wind.

It is the voice of the Earth, calling not to reprimand but to remind. She speaks to you, her child, not with anger but with hope. She reminds you that her story is your story, that her health is your future, and that her balance is your legacy.

'Do you remember me?' she asks gently, her voice as ancient as time itself. 'I cradled your ancestors, offered them shelter, and nourished them with my gifts. I have weathered storms and carried civilisations through the rise and fall of ages. I have given endlessly—my rivers, my forests, my soil. Yet now, I find myself weary, burdened by the weight of your progress. And still, I whisper to you, because I believe in the goodness that lies within you.'

Her whisper is not a demand but an invitation—a call to reconnect with the essence of who you are. At its core, the Earth's message is about balance: the harmony between ambition and humility, between progress and preservation, and between what you take and what you give back.

The Earth knows that the imbalance we see in the world today mirrors an imbalance within ourselves. When your days are filled with rush and distraction, when you measure success by what you own rather than what you nurture, the disconnection grows. Yet, the Earth whispers to remind you that true harmony begins within.

'Look inward, my child,' she says, 'for the peace you seek in the world begins in your heart. When you live with integrity, compassion, and mindfulness, you create ripples that extend far beyond yourself. When your actions are guided by kindness and care, the world around you begins to heal.'

Her wisdom is timeless, rooted in stories passed down through generations. The myths and teachings of your ancestors all point to one truth: balance is the key to thriving. The dance of Lord Nataraja, the selfless acts of King Prithu, and the enduring lessons of Prometheus, all serve as reminders that progress and sustainability are not opposing forces but partners in the eternal dance of life.

To be a good citizen of this Earth is to embody values that uplift not just yourself but the community of life you share with every living being. Compassion, responsibility, and humility are the cornerstones of this citizenship.

When you choose to act with compassion, you become a steward of life itself. When you take responsibility for your choices—what you consume, how you live, and the legacy you leave—you honour the sacred trust between humanity and nature. And when you embrace humility, recognising that you are not the master but the custodian of this planet, you open the door to true progress—progress that uplifts without destroying, that builds without depleting.

The Earth's whisper carries this plea: 'Live not as conquerors, but as caretakers. Take only what you need, and give back in ways that nurture the web of life. Plant seeds—not just in the soil, but in the hearts of those around you. Let your actions inspire others to listen to my whisper, to remember what we have forgotten, and to restore the balance we so desperately need.'

Imagine a world where every action you take is guided by the Earth's whisper—a world where cities are built to coexist with nature, where rivers run clean, and where forests stand tall, offering sanctuary to countless forms of life. This vision is not a distant dream; it is a choice that begins with you.

The Earth does not ask for perfection. She does not demand grand gestures or impossible sacrifices. All she asks is that you listen and act with intention. Choose to walk lightly upon her soil. Choose to see her rivers as sacred, her forests as family, and her creatures as kin.

As you move through your days, let the Earth's whisper be your guide. When faced with a choice, ask yourself: Does this bring balance or disrupt it? Does this nourish or deplete? Does this create harmony or sow discord?

'My child,' the Earth says, her voice tender, yet strong, 'you are my hope. Your choices today shape the world of tomorrow. Listen to my whispers, not as a burden but as a gift—a chance to leave a legacy of healing and harmony.

Together, we can create a world where my voice is no

longer a lament but a song, where every breath you take is a testament to the balance we have restored.'

Let her whisper guide you. Let it awaken the values that define a good citizen—compassion, responsibility, and a deep respect for the interconnectedness of all life. Let it remind you that you are not separate from the Earth; you are a part of her, and her whispers are the echoes of your own soul.

And so, the Earth whispers once more, her voice carrying through the wind and the water, through the fire and the soil: 'Will you listen?'